Atmel AVR Microcontroller Primer: Programming and Interfacing

Copyright © 2008 by Morgan & Claypool

All rights reserved. No part of this publication may be reproduced, stored in a retrieval system, or transmitted in any form or by any means—electronic, mechanical, photocopy, recording, or any other except for brief quotations in printed reviews, without the prior permission of the publisher.

Atmel AVR Microcontroller Primer: Programming and Interfacing
Steven F. Barrett and Daniel J. Pack
www.morganclaypool.com

ISBN: 1598295411   paperback
ISBN: 9781598295412   paperback

ISBN: 159829542X   ebook
ISBN: 9781598295429   ebook

DOI: 10.2200/S00100ED1V01Y200712DCS015
A Publication in the Morgan & Claypool Publishers series

SYNTHESIS LECTURES ON DIGITAL CIRCUITS AND SYSTEMS # 15

Lecture #15
Series Editor: Mitchell A. Thornton, Southern Methodist University

Series ISSN

ISSN 1932-3166   print
ISSN 1932-3174   electronic

Atmel AVR Microcontroller Primer: Programming and Interfacing

Steven F. Barrett
University of Wyoming

Daniel J. Pack
United States Air Force Academy

SYNTHESIS LECTURES ON DIGITAL CIRCUITS AND SYSTEMS #15

ABSTRACT

This textbook provides practicing scientists and engineers a primer on the Atmel AVR microcontroller. Our approach is to provide the fundamental skills to quickly get up and operating with this internationally popular microcontroller. The Atmel ATmega16 is used as a representative sample of the AVR line. The knowledge you gain on the ATmega16 can be easily translated to every other microcontroller in the AVR line. We cover the main subsystems aboard the ATmega16, providing a short theory section followed by a description of the related microcontroller subsystem with accompanying hardware and software to exercise the subsytem. In all examples, we use the C programming language. We conclude with a detailed chapter describing how to interface the microcontroller to a wide variety of input and output devices.

KEYWORDS

Atmel microcontroller, microcontroller, ATmega16, Atmel AVR, microcontroller interfacing

Preface

In 2006, Morgan & Claypool Publishers (M&C) released our textbook *Microcontrollers Fundamentals for Engineers and Scientists*. The purpose of this textbook was to provide practicing scientists and engineers a tutorial on the fundamental concepts and the use of microcontrollers. The textbook presented the fundamental concepts common to all microcontrollers. Our goals for writing this follow-on book are to present details on a specific microcontroller family—the Atmel AVR Microcontroller.

Why Atmel? There are many excellent international companies that produce microcontrollers. As Atmel states, "Atmel Corporation is an industry leader in the design and manufacture of advanced semiconductors, with focus on microcontrollers, nonvolatile memory, logic, radio frequency components and sensors." Some of the highlights of the Atmel AVR line include

- high performance coupled with low power consumption,
- outstanding flash memory technology,
- reduced instruction set computer Harvard Architecture,
- single-cycle instruction execution,
- wide variety of operating voltages (1.8–5.5 VDC),
- architecture designed for the C language,
- one set of development tools for the entire AVR line, and
- in-system programming, debugging, and verification capability.

Although all of these features are extremely important, we have chosen to focus on the Atmel AVR line of microcontrollers for this primer for a number of other related reasons:

- The learning curve for Atmel microcontrollers is gentle. If this is your first exposure to microcontrollers, you will quickly come up to speed on microcontroller programming and interfacing. If you already know another line of processors, you can quickly apply your knowledge to this powerful line of 8-bit processors.
- It is relatively inexpensive to get started with the Atmel AVR microcontroller line. The microcontrollers themselves are inexpensive, and the compilers and programming hardware and software are relatively inexpensive.

- The AVR line provides a full range of processing power, from small 8-pin processors to complex 100-pin processors. The same compiler and programming hardware may be used with a wide variety of microcontrollers.
- Many of the AVR microcontrollers are available in dual inline package, which makes them readily useable on a printed circuit board prototype (e.g., senior design projects).
- Many of the microcontrollers in the AVR line are pin-for-pin compatible with one another. This allows you to easily move up and down the AVR line as your project becomes better defined.
- Atmel has documentation available for every microcontroller at your fingertips. Simply visit www.atmel.com. Furthermore, Atmel customer support is good and responsive.
- There is worldwide interest in the AVR microcontroller line. We would be remiss to not mention AVR Freaks. This is a dedicated, international group of AVR experts who share their expertise online with other high-power users and novices alike.

Approach of the book

If this is your first exposure to microcontrollers, we highly recommend that you read first our other M&C textbook, *Microcontrollers Fundamentals for Engineers and Scientists*. It will provide you the background information necessary to fully appreciate the contents of this textbook. This textbook picks up where the first one left off. We have received permission from M&C to include some of the background material from the first textbook in this text to allow for a complete stand-alone product.

Our approach in this textbook is to provide you the fundamental skills to quickly get up and operating with an Atmel microcontroller. We have chosen to use the AVR ATmega16 as a representative sample of the AVR line (more on this processor later). The knowledge you gain on the ATmega16 can be easily translated to every other microcontroller in the AVR line.

We will use an ongoing testbench example throughout the textbook. We will start by having you get a simple microcontroller circuit operating with a simple menu program that interacts with external devices. As we move through various microcontroller subsystems, we will continue to add features to the testbench. By the end of the textbook, you will have a complete hardware/software system that demonstrates the features of the ATmega16. You can then use this testbench to adapt to other applications.

The M&C textbooks are designed to be short tutorials on a given topic. Therefore, our treatment of each topic will provide a short theory section followed by a description of the related microcontroller subsystem with accompanying hardware and software to exercise the subsystem. In all examples, we will use the C programming language. There are many excellent C compilers available for the Atmel AVR line. We have chosen the ImageCraft ICC AVR compiler for its short learning curve and ease of use.

Acknowledgments

Space does not permit us to thank everyone who has provided encouragement along the way. We thank Joel Claypool and John Enderle for inviting us to participate in their efforts to develop a series of short tutorial textbooks on select engineering topics. We also thank Atmel and ImageCraft for their permission to use their copyrighted material and screenshots throughout the text. We especially thank Helen Perlegos of Atmel for her assistance is securing appropriate permission to use Atmel material within the text.

Most of all, we thank our families. We acknowledge our parents. Thank you, Moms (Eleanore and Jackie), and thank you, Dad (Frank), for always believing in me (S.B.). Thank you, Moms (Young Shin and Rana), and thank you, Dads (Sung Bock and Chong Kon), for your encouragement and unfailing support (D.P.). Finally, our work could not have come to fruition without the sacrifices of our family members: Cindy, Heidi, Heather, Jon R., Christine, Jon B., Andrew, and Graham. As always, without you none of this would matter. We love you!

Laramie and Colorado Springs, November 2007

Steve Barrett and Daniel Pack

Contents

1. **Atmel AVR Architecture Overview** 1
   1.1 ATmega16 Architecture Overview 1
      1.1.1 Reduced Instruction Set Computer 1
      1.1.2 Assembly Language Instruction Set 2
      1.1.3 ATmega16 Architecture Overview 3
   1.2 Nonvolatile and Data Memories 3
      1.2.1 In-System Programmable Flash EEPROM 3
      1.2.2 Byte-Addressable EEPROM 5
      1.2.3 Static Random Access Memory 5
      1.2.4 Programmable Lock Bits 5
   1.3 Port System 6
   1.4 Peripheral Features—Internal Subsystems 8
      1.4.1 Time Base 8
      1.4.2 Timing Subsystem 9
      1.4.3 Pulse Width Modulation Channels 9
      1.4.4 Serial Communications 9
         1.4.4.1 Serial USART 9
         1.4.4.2 Serial Peripheral Interface 10
         1.4.4.3 Two-Wire Serial Interface 10
      1.4.5 Analog-to-Digital Converter 10
      1.4.6 Interrupts 11
   1.5 Physical and Operating Parameters 11
      1.5.1 Packaging 11
      1.5.2 Power Consumption 11
      1.5.3 Speed Grades 13
   1.6 Application: ATmega16 Testbench 13
      1.6.1 Hardware Configuration 13
      1.6.2 Software Configuration 15

    1.7 Programming the ATmega16 19
        1.7.1   Programming Procedure 20
    1.8 Software Portability 22
    1.9 Summary 23
    1.10 References and Further Reading 23
    1.11 Chapter Problems 23

2. Serial Communication Subsystem 25
    2.1 Serial Communications 25
    2.2 Serial Communication Terminology 25
        2.2.1   Asynchronous versus Synchronous Serial Transmission 26
        2.2.2   Baud Rate 26
        2.2.3   Full Duplex 26
        2.2.4   Nonreturn to Zero Coding Format 26
        2.2.5   The RS-232 Communication Protocol 27
        2.2.6   Parity 27
        2.2.7   American Standard Code for Information Interchange 27
    2.3 Serial USART 27
        2.3.1   System Overview 28
            2.3.1.1   USART Clock Generator 29
            2.3.1.2   USART Transmitter 30
            2.3.1.3   USART Receiver 30
            2.3.1.4   USART Registers 30
        2.3.2   System Operation and Programming 32
        2.3.3   Serial Peripheral Interface 34
            2.3.3.1   SPI Operation 34
            2.3.3.2   Registers 35
            2.3.3.3   Programming 37
    2.4 Two-Wire Serial Interface 38
    2.5 Summary 38
    2.6 References and Further Reading 38
    2.7 Chapter Problems 39

3. Analog-to-Digital Conversion 41
    3.1 Background Theory 41
        3.1.1   Analog versus Digital Signals 42
        3.1.2   Sampling, Quantization, and Encoding 44
        3.1.3   Resolution and Data Rate 48

3.2	Analog-To-Digital Conversion Process	50
3.3	ADC Conversion Technologies	53
	3.3.1 Successive Approximation	53
	3.3.2 Integration	55
	3.3.3 Counter-Based Conversion	55
	3.3.4 Parallel Conversion	55
3.4	The Atmel ATmega16 ADC System	55
	3.4.1 Block Diagram	56
	3.4.2 Registers	58
	3.4.2.1 ADC Multiplexer Selection Register	58
	3.4.2.2 ADC Control and Status Register A	59
	3.4.2.3 ADC Data Registers (ADCH and ADCL)	59
	3.4.3 Programming the ADC	59
	3.4.4 Digital-to-Analog Conversion	62
3.5	Summary	63
3.6	References and Further Reading	63
3.7	Chapter Problems	64

4. Interrupt Subsystem 65

4.1	Interrupt Theory	65
4.2	ATmega16 Interrupt System	65
4.3	Programming An Interrupt	66
4.4	Application	68
	4.4.1 External Interrupts	68
	4.4.2 Internal Interrupt	71
4.5	Summary	74
4.6	References and Further Reading	74
4.7	Chapter Problems	74

5. Timing Subsystem 75

5.1	Overview	75
5.2	Timing-Related Terminology	76
	5.2.1 Frequency	76
	5.2.2 Period	76
	5.2.3 Duty Cycle	76
5.3	Timing System Overview	76
5.4	Applications	79
	5.4.1 Input Capture—Measuring External Timing Event	79

		5.4.2	Counting Events..81

- 5.4.2 Counting Events ... 81
- 5.4.3 Output Compare—Generating Timing Signals to Interface External Devices ... 81
- 5.4.4 Industrial Implementation Case Study (PWM) ... 82
- 5.5 Overview of the Atmel Timers ... 83
- 5.6 Timer 0 System ... 84
  - 5.6.1 Modes of Operation ... 86
    - 5.6.1.1 Normal Mode ... 87
    - 5.6.1.2 Clear Timer on Compare Match ... 87
    - 5.6.1.3 Phase Correct PWM Mode ... 87
    - 5.6.1.4 Fast PWM ... 87
  - 5.6.2 Timer 0 Registers ... 87
    - 5.6.2.1 Timer/Counter Control Register 0 ... 88
    - 5.6.2.2 Timer/Counter Register ... 88
    - 5.6.2.3 Output Compare Register ... 88
    - 5.6.2.4 Timer/Counter Interrupt Mask Register ... 90
    - 5.6.2.5 Timer/Counter Interrupt Flag Register ... 91
- 5.7 Timer 1 ... 91
  - 5.7.1 Timer 1 Registers ... 91
    - 5.7.1.1 TCCR1A and TCCR1B Registers ... 91
    - 5.7.1.2 Timer/Counter Register 1 (TCNT1H/TCNT1) ... 91
    - 5.7.1.3 Output Compare Register 1 Channel A (OCR1AH/OCR1AL) ... 94
    - 5.7.1.4 Output Compare Register 1 Channel B (OCR1BH/OCR1BL) ... 94
    - 5.7.1.5 Input Capture Register 1 (ICR1H/ICR1L) ... 94
    - 5.7.1.6 Timer/Counter Interrupt Mask Register (TIMSK) ... 94
    - 5.7.1.7 Timer/Counter Interrupt Flag Register (TIFR) ... 94
- 5.8 Timer 2 ... 94
  - 5.8.1 Timer/Counter Control Register 2 ... 94
  - 5.8.2 Timer/Counter Register (TCNT2) ... 95
  - 5.8.3 Output Compare Register (OCR2) ... 95
  - 5.8.4 Timer/Counter Interrupt Mask Register (TIMSK) ... 96
  - 5.8.5 Timer/Counter Interrupt Flag Register ... 96
- 5.9 Programming the Timer System ... 96
  - 5.9.1 Precision Delay ... 98

| | | 5.9.2 | Pulse Width Modulation .... 99 |
| | | 5.9.3 | Input Capture Mode .... 101 |

    5.10 Summary 103
    5.11 References and Further Reading 103
    5.12 Chapter Problems 104

6. Atmel AVR Operating Parameters and Interfacing **105**
    6.1 Operating Parameters 106
    6.2 Input Devices 107
        6.2.1 Switches 109
        6.2.2 Switch Debouncing 110
        6.2.3 Keypads 111
        6.2.4 Sensors 111
            6.2.4.1 Digital Sensors 111
            6.2.4.2 Analog Sensors 114
    6.3 Output Devices 114
        6.3.1 Light-Emitting Diodes 115
        6.3.2 Seven-Segment LED Displays 117
        6.3.3 Tristate LED Indicator 117
        6.3.4 Dot Matrix Display 120
        6.3.5 Liquid Crystal Display 120
        6.3.6 High-Power DC Devices 124
    6.4 DC Motor Speed and Direction Control 125
        6.4.1 DC Motor Operating Parameters 126
        6.4.2 AC Devices 126
    6.5 Application: Flight Simulator Panel 127
    6.6 Summary 157
    6.7 References and Further Reading 158
    6.8 Chapter Problems 158

A. ATmega16 Register Set **159**
B. ATmega16 Header File **161**
    Author Biography 177
    Index 179

CHAPTER 1

Atmel AVR Architecture Overview

Objectives: After reading this chapter, the reader should be able to

- provide an overview of the RISC architecture of the ATmega16,
- describe the different ATmega16 memory components and their applications,
- explain the ATmega16 internal subsystems and their applications,
- highlight the operating parameters of the ATmega16, and
- summarize the special ATmega16 features.

1.1 ATmega16 ARCHITECTURE OVERVIEW

In this section, we describe the overall architecture of the Atmel AVR ATmega16. We begin with an introduction to the concept of the reduced instruction set computer (RISC) and briefly describe the Atmel Assembly Language Instruction Set. A brief introduction is warranted because we will be programming mainly in C throughout the course of the book. We then provide a detailed description of the ATmega16 hardware architecture.

1.1.1 Reduced Instruction Set Computer

In our first Morgan & Claypool (M&C) [1] textbook, we described a microcontroller as an entire computer system contained within a single integrated circuit or chip. Microcontroller operation is controlled by a user-written program interacting with the fixed hardware architecture resident within the microcontroller. A specific microcontroller architecture can be categorized as accumulator-based, register-based, stack-based, or a pipeline architecture.

The Atmel ATmega16 is a register-based architecture. In this type of architecture, both operands of an operation are stored in registers collocated with the central processing unit (CPU). This means that before an operation is performed, the computer loads all necessary data for the operation to its CPU. The result of the operation is also stored in a register. During program execution, the CPU interacts with the register set and minimizes slower memory accesses. Memory accesses are typically handled as background operations.

Coupled with the register-based architecture is an instruction set based on the RISC concept. A RISC processor is equipped with a complement of very simple and efficient basic operations. More complex instructions are built up from these very basic operations. This allows for efficient program operation. The Atmel ATmega16 is equipped with 131 RISC-type instructions. Most can be executed in a single clock cycle. The ATmega16 is also equipped with additional hardware to allow for the multiplication operation in two clock cycles. In many other microcontroller architectures, multiplication typically requires many more clock cycles. For additional information on the RISC architecture, the interested reader is referred to Hennessy and Patterson [3].

The Atmel ATmega16 [2] is equipped with 32 general purpose 8-bit registers that are tightly coupled to the processor's arithmetic logic unit within the CPU. Also, the processor is designed following the Harvard Architecture format. That is, it is equipped with separate, dedicated memories and buses for program and data information. The register-based Harvard Architecture coupled with the RISC-based instruction set allows for fast and efficient program execution and allows the processor to complete an assembly language instruction every clock cycle. Atmel indicates the ATmega16 can execute 16 million instructions per second when operating at a clock speed of 16 MHz.

1.1.2 Assembly Language Instruction Set

An instruction set is a group of instructions a machine "understands" to execute. A large number of instructions provide flexibility but require more complex hardware. Thus, an instruction set is unique for a given hardware and cannot be used with another hardware configuration. Atmel has equipped the ATmega16 with 131 different instructions.

For the most efficient and fast execution of a given microcontroller, assembly language should be used. Assembly language is written to efficiently interact with a specific microcontroller's resident hardware. To effectively use the assembly language, the programmer must be thoroughly familiar with the low-level architecture details of the controller. Furthermore, the learning curve for a given assembly language is quite steep and lessons learned do not always transfer to another microcontroller.

We will program the Atmel ATmega16 using the C language throughout the text. The C programming language allows for direct control of microcontroller hardware at the register level while being portable to other microcontrollers in the AVR line. When a C program is compiled during the software development process, the program is first converted to assembly language and then to the machine code for the specific microcontroller.

We must emphasize that programming in C is not better than assembly language or vice versa. Both approaches have their inherent advantages and disadvantages. We have chosen to use C in this textbook for the reasons previously discussed.

1.1.3 ATmega16 Architecture Overview

We have chosen the ATmega16 as a representative of the Atmel AVR line of microcontrollers. Lessons learned with the ATmega16 may be easily adapted to all other processors in the AVR line. A block diagram of the Atmel ATmega16's architecture is provided in Figure 1.1.

As can be seen from the figure, the ATmega16 has external connections for power supplies (VCC, GND, AVCC, and AREF), an external time base (XTAL1 and XTAL2) input pins to drive its clocks, processor reset (active low RESET), and four 8-bit ports (PA0-PA7, PC0-PC7, PB0-PB7, and PD0-PD7), which are used to interact with the external world. As we shall soon see, these ports may be used as general purpose digital input/output (I/O) ports or they may be used for the alternate functions. The ports are interconnected with the ATmega16's CPU and internal subsystems via an internal bus. The ATmega16 also contains a timer subsystem, an analog-to-digital converter (ADC), an interrupt subsystem, memory components, and a communication subsystem.

In the next several subsections, we briefly describe each of these internal subsystems shown in the figure. Detailed descriptions of selected subsystem operation and programming are provided later in this book. We cannot cover all features of the microcontroller because of limited space. Instead, we focus on the primary functional components of the ATmega16 to fulfill the purpose of this book as a basic primer to the ATmega16.

1.2 NONVOLATILE AND DATA MEMORIES

The ATmega16 is equipped with three main memory sections: flash electrically erasable programmable read-only memory (EEPROM), static random access memory (SRAM), and byte-addressable EEPROM for data storage. We discuss each memory component in turn.

1.2.1 In-System Programmable Flash EEPROM

Bulk programmable flash EEPROM is used to store programs. It can be erased and programmed as a single unit. Also, should a program require a large table of constants, it may be included as a global variable within a program and programmed into flash EEPROM with the rest of the program. Flash EEPROM is nonvolatile, meaning memory contents are retained when microcontroller power is lost. The ATmega16 is equipped with 16K bytes of onboard reprogrammable flash memory. This memory component is organized into 8K locations, with 16 bits at each location.

The flash EEPROM is in-system programmable. In-system programmability means the microcontroller can be programmed while resident within a circuit. It does not have to be removed from the circuit for programming. Instead, a host personal computer (PC) connected via a cable to a microcontroller downloads the program to the microcontroller. Alternately, the microcontroller

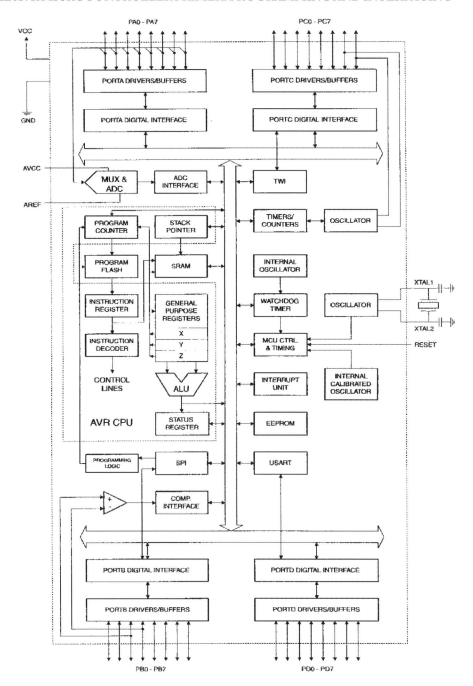

FIGURE 1.1: Atmel AVR ATmega16 block diagram. Figure used with permission of Atmel.

can be programmed outside its resident circuit using a flash programmer board. We will use this technique throughout the book. Specifically, we will use the Atmel STK500 AVR Flash MCU Starter Kit for programming the ATmega16. This inexpensive development board (less than $100) is readily available from a number of suppliers.

1.2.2 Byte-Addressable EEPROM

Byte-addressable memory is used to permanently store and recall variables during program execution. It too is nonvolatile. It is especially useful for logging system malfunctions and fault data during program execution. It is also useful for storing data that must be retained during a power failure but might need to be changed periodically. Examples where this type of memory is used are found in applications to store system parameters, electronic lock combinations, and automatic garage door electronic unlock sequences. The ATmega16 is equipped with 512 bytes of EEPROM.

1.2.3 Static Random Access Memory

SRAM is volatile. That is, if the microcontroller loses power, the contents of SRAM memory are lost. It can be written to and read from during program execution. The ATmega16 is equipped with 1000 bytes (actually 1120) of SRAM. A small portion (96 locations) of the SRAM is set aside for the general-purpose registers used by the CPU and also for the I/O and peripheral subsystems aboard the microcontroller. A complete ATmega16 register listing and accompanying header file is provided in Appendices A and B, respectively. During program execution, RAM is used to store global variables, support dynamic memory allocation of variables, and provide a location for the stack (to be discussed later).

1.2.4 Programmable Lock Bits

To provide for memory security from tampering, the ATmega16 is equipped with six memory lock bits. These lock bits are programmed using the Atmel STK500 programming board. The lock bits may be configured for the following options:
- No memory lock features enabled.
- No further programming of memory is allowed using parallel or serial programming techniques.
- No further programming or verification of memory is allowed using parallel or serial programming techniques.

6 ATMEL AVR MICROCONTROLLER PRIMER: PROGRAMMING AND INTERFACING

1.3 PORT SYSTEM

The Atmel ATmega16 is equipped with four 8-bit general-purpose, digital I/O ports designated PORTA, PORTB, PORTC, and PORTD. All of these ports also have alternate functions, which will be described later. In this section, we concentrate on the basic digital I/O port features.

As shown in Figure 1.2, each port has three registers associated with it:

- Data Register (PORTx)—used to write output data to the port,

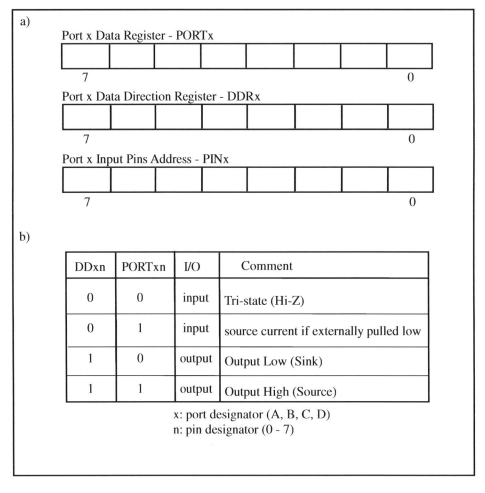

FIGURE 1.2: ATmega16 port configuration registers: (a) port-associated registers and (b) port pin configuration.

- Data Direction Register (DDRx)—used to set a specific port pin to either output (1) or input (0), and
- Input Pin Address (PINx)—used to read input data from the port.

Figure 1.2(b) describes the settings required to configure a specific port pin to either input or output. If selected for input, the pin may be selected for either an input pin or to operate in the high-impedance (Hi-Z) mode. In Hi-Z mode, the input appears as high impedance to a particular pin. If selected for output, the pin may be further configured for either logic low or logic high.

Port pins are usually configured at the beginning of a program for either input or output, and their initial values are then set. Usually, all eight pins for a given port are configured simultaneously. A code example is provided below to show how ports are configured. Note that because we are using the C programming language with a compiler include file, the register contents are simply referred to by name. Note that the data direction register (DDRx) is first used to set the pins as either input or output, and then the data register (PORTx) is used to set the initial value of the output port pins.

```
//***************************************************************
//initialize_ports: provides initial configuration for I/O ports
//***************************************************************

void initialize_ports(void)
{
DDRA=0xfc;   //set PORTA[7:2] as output, PORTA[1:0]
             //as input (1111_1100)
PORTA=0x03;  //initialize PORTA[7:2] low, PORTA[1:0]
             //current source
DDRB=0xa0;   //PORTB[7:4] as output, set PORTB[3:0] as input
PORTB=0x00;  //disable PORTB pull-up resistors


DDRC=0xff;   //set PORTC as output
PORTC=0x00;  //initialize low
DDRD=0xff;   //set PORTD as output
PORTD=0x00;  //initialize low
}
```

To read the value from a port pin configured as input, the following code could be used. Note the variable used to read the value from the input pins is declared as an unsigned char because both the port and this variable type are 8 bits wide.

```
unsigned char   new_PORTB;              //new values of PORTB
   :
   :
   :
new_PORTB = PINB;                       //read PORTB
```

1.4 PERIPHERAL FEATURES—INTERNAL SUBSYSTEMS

In this section, we provide a brief overview of the peripheral features of the ATmega16. It should be emphasized that these features are the internal subsystems contained within the confines of the microcontroller chip. These built-in features allow complex and sophisticated tasks to be accomplished by the microcontroller.

1.4.1 Time Base

The microcontroller is a complex synchronous state machine. It responds to program steps in a sequential manner as dictated by a user-written program. The microcontroller sequences through a predictable fetch–decode–execute sequence. Each unique assembly language program instruction issues a series of signals to control the microcontroller hardware to accomplish instruction related operations.

The speed at which a microcontroller sequences through these actions is controlled by a precise time base called the clock. The clock source is routed throughout the microcontroller to provide a time base for all peripheral subsystems. The ATmega16 may be clocked internally, using a user-selectable resistor capacitor (RC) time base, or externally. The RC internal time base is selected using programmable fuse bits. We will discuss how to do this in the application section of this chapter. You may choose an internal fixed clock operating frequency of 1, 2, 4, or 8 MHz.

To provide for a wider range of frequency selections, an external time source may be used. The external time sources, in order of increasing accuracy and stability, are an external RC network, a ceramic resonator, or a crystal oscillator. The system designer chooses the time base frequency and clock source device appropriate for the application at hand.

1.4.2 Timing Subsystem

The ATmega16 is equipped with a complement of timers that allows the user to generate a precision output signal, measure the characteristics (period, duty cycle, frequency) of an incoming digital signal, or count external events. Specifically, the ATmega16 is equipped with two 8-bit timer/counters and one 16-bit counter. We discuss the operation, programming, and application of the timing system in Chapter 5 of the text.

1.4.3 Pulse Width Modulation Channels

A pulse width modulated, or PWM, signal is characterized by a fixed frequency and a varying duty cycle. Duty cycle is the percentage of time a repetitive signal is logic high during the signal period. It may be formally expressed as

$$\text{duty cycle } (\%) = (\text{on time}/\text{period}) \times (100\%).$$

The ATmega16 is equipped with four PWM channels. The PWM channels coupled with the flexibility of dividing the time base down to different PWM subsystem clock source frequencies allows the user to generate a wide variety of PWM signals, from relatively high-frequency, low-duty cycle signals to relatively low-frequency, high-duty cycle signals.

PWM signals are used in a wide variety of applications, including controlling the position of a servo motor and controlling the speed of a DC motor. We discuss the operation, programming, and application of the PWM system in Chapter 5 of the text.

1.4.4 Serial Communications

The ATmega16 is equipped with a host of different serial communication subsystems, including the Universal Synchronous and Asynchronous Serial Receiver and Transmitter (USART), the Serial Peripheral Interface (SPI), and the Two-Wire Serial Interface (TWI). What all of these systems have in common is the serial transmission of data. In a serial communications transmission scheme, data are sent a single bit at a time from transmitter to receiver.

1.4.4.1 Serial USART. The serial USART is used for full duplex (two-way) communication between a receiver and transmitter. This is accomplished by equipping the ATmega16 with independent hardware for the transmitter and receiver. The USART is typically used for asynchronous communication. That is, there is not a common clock between the transmitter and receiver to keep them synchronized with one another. To maintain synchronization between the transmitter and receiver, framing start and stop bits are used at the beginning and end of each data byte in a transmission sequence.

The ATmega16 USART is quite flexible. It has the capability to be set to a variety of data transmission rates known as the baud (bits per second) rate. The USART may also be set for data bit widths of 5 to 9 bits with one or two stop bits. Furthermore, the ATmega16 is equipped with a hardware-generated parity bit (even or odd) and parity check hardware at the receiver. A single parity bit allows for the detection of a single bit error within a byte of data. The USART may also be configured to operate in a synchronous mode. We discuss the operation, programming, and application of the USART in Chapter 2 of the text.

1.4.4.2 Serial Peripheral Interface. The ATmega16 SPI can also be used for two-way serial communication between a transmitter and a receiver. In the SPI system, the transmitter and receiver share a common clock source. This requires an additional clock line between the transmitter and receiver but allows for higher data transmission rates as compared with the USART.

The SPI may be viewed as a synchronous 16-bit shift register with an 8-bit half residing in the transmitter and the other 8-bit half residing in the receiver. The transmitter is designated the master because it provides the synchronizing clock source between the transmitter and the receiver. The receiver is designated as the slave. We discuss the operation, programming, and application of the SPI in Chapter 2 of the text.

1.4.4.3 Two-Wire Serial Interface. The TWI subsystem allows the system designer to network a number of related devices (microcontrollers, transducers, displays, memory storage, etc.) together into a system using a two-wire interconnecting scheme. The TWI allows a maximum of 128 devices to be connected together. Each device has its own unique address and may both transmit and receive over the two-wire bus at frequencies up to 400 kHz. This allows the device to freely exchange information with other devices in the network within a small area.

1.4.5 Analog-to-Digital Converter

The ATmega16 is equipped with an eight-channel ADC subsystem. The ADC converts an analog signal from the outside world into a binary representation suitable for use by the microcontroller. The ATmega16 ADC has 10-bit resolution. This means that an analog voltage between 0 and 5 V will be encoded into one of 1024 binary representations between $(000)_{16}$ and $(3FF)_{16}$. This provides the ATmega16 with a voltage resolution of approximately 4.88 mV. We discuss the operation, programming, and application of the ADC in Chapter 3 of the text.

1.4.6 Interrupts

The normal execution of a program step follows a designated sequence of instructions. However, sometimes, this normal sequence of events must be interrupted to respond to high-priority faults and status both inside and outside the microcontroller. When these higher-priority events occur, the microcontroller must temporarily suspend normal operation and execute event specific actions called an interrupt service routine. Once the higher priority event has been serviced, the microcontroller returns and continues processing the normal program.

The ATmega16 is equipped with a complement of 21 interrupt sources. Three of the interrupts are provided for external interrupt sources, whereas the remaining 19 interrupts support the efficient operation of peripheral subsystems aboard the microcontroller. We discuss the operation, programming, and application of the interrupt system in Chapter 4 of the text.

1.5 PHYSICAL AND OPERATING PARAMETERS

In this section, we provide data on the physical layout and operating parameters of the ATmega16 microcontroller. As a system designer, it is important to know the various physical and operating parameter options available to select the best option for a given application.

1.5.1 Packaging

The ATmega16 comes in three different packaging styles: a 40-pin plastic dual in-line package (DIP), a 44-lead thin quad flat pack package, and a 44-pad quad flat nonlead/microlead frame package. The Pinout Diagram for the different packaging options are provided in Figure 1.3.

1.5.2 Power Consumption

The ATmega16 is available at two different operating voltage ranges. The ATmega16L operates at supply voltages from 2.7 to 5.5 VDC, whereas the ATmega16 operates at supply voltages from 4.5 to 5.5 VDC. In the application examples that follow, we will use a standard laboratory 5-VDC power supply and also discuss a method of providing a 5-VDC supply using an off-the-shelf 9-VDC battery.

The current draw for the microcontroller is quite low. For example, when the ATmega16L is actively operating at 3 MHz from a 3-VDC power source, the current draw is 1.1 mA. When placed in the idle mode, the microcontroller current draw reduces to less than 0.35 mA. Finally, in the power-down mode, the microcontroller will draw less than 1 μA of current from the voltage source.

To minimize power consumption, the microcontroller can be placed into various low-current sleep modes. There are six different sleep modes available to the system designer. The

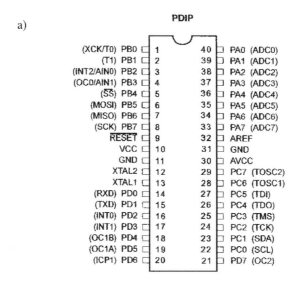

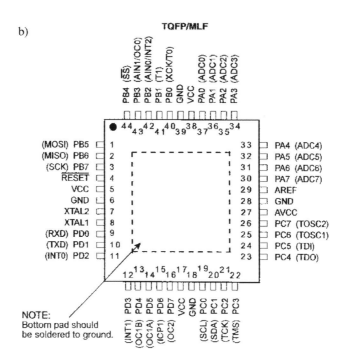

FIGURE 1.3: Atmel AVR ATmega16 Pinout Diagram: (a) 40-pin plastic DIP and (b) thin quad flat pack/microlead frame. Figure used with permission of Atmel.

microcontroller is placed in sleep mode using the SLEEP command and "wakened" from SLEEP when an interrupt occurs. Additionally, power consumption can be further reduced by operating the microcontroller at the lowest practical clock frequency for a given application.

1.5.3 Speed Grades

The ATmega16 is available at two different speed grades. The ATmega16L operates from 0 to 8 MHz, whereas the ATmega16 operates from 0 to 16 MHz. As previously discussed, the operating speed of the microcontroller is set by the time base chosen for the processor. One might believe that faster microcontroller operation is always better. This is not the case. The system designer must determine the minimum practical speed of microcontroller operation for a given application. The microcontroller's power consumption is directly related to operating speed. That is, the faster the operating speed of the microcontroller, the higher its power consumption. This becomes especially critical in portable, battery-operated embedded systems applications.

That completes our brief introduction to the features of the ATmega16. In the next section, we apply what we have learned in developing a testbench for the ATmega16.

1.6 APPLICATION: ATmega16 TESTBENCH

In the application sections throughout the text, we use a running example—an ATmega16 Testbench. The purpose of the Testbench is to illustrate the operation of selected ATmega16 subsystems working with various I/O devices. Most importantly, the Testbench will serve as a template to develop your own applications.

In this chapter, we provide the hardware configuration of a barebones testbench and a basic software framework to get the system up and operating. We will connect eight debounced tact switches to PORTB and an eight-channel tristate light-emitting diode (LED) array to PORTA. The software will check for a status change on PORTB. When the user depresses one of the tact switches, the ATmega16 will detect the status change and the corresponding LED on PORTA will transition from red to green.

1.6.1 Hardware Configuration

Provided in Figure 1.4 is the basic hardware configuration for the Testbench. We will discuss in detail the operation of the I/O devices in Chapter 6.

PORTA is configured with eight tact (momentary) switches with accompanying debouncing hardware. We discuss the debounce circuit in detail in Chapter 6. PORTB is equipped with an eight-channel tristate LED indicator. For a given port pin, the green LED will illuminate for a logic high, the red LED for a logic low, and no LEDs for a tristate high-impedance state. We discuss this circuit in detail in Chapter 6.

14 ATMEL AVR MICROCONTROLLER PRIMER: PROGRAMMING AND INTERFACING

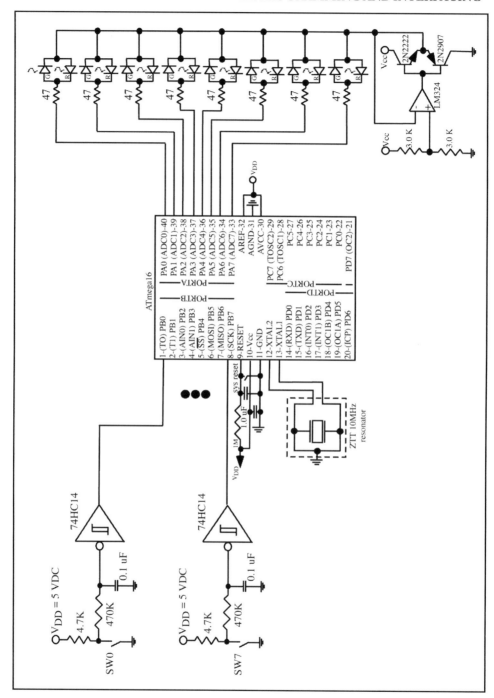

FIGURE 1.4: ATmega16 Testbench hardware.

Aside from the input hardware on PORTB and the output display hardware on PORTA of the controller, there are power (pins 10, 30, and 32) and ground (pins 11 and 31) connections. A standard 5-VDC power supply may be used for the power connections. For portable applications, a 9-VDC battery equipped with a 5-VDC regulator (LM340-05 or uA7805) may be used as a power source. Pins 9 through 11 have a resistor (1 M), two capacitors (1.0 μF), and a tact switch configured to provide a reset switch for the microcontroller. We use a ZTT 10-MHz ceramic resonator as the time base for the Testbench. It is connected to pins 12 (XTAL2) and 13 (XTAL1) of the ATmega16.

1.6.2 Software Configuration

The Testbench software is provided below. The program contains the following sections:

- Comments
- Include Files: We have included the ImageCraft ICC AVR include file for the ATmega16 (iom16v.h). This file provides the software link between the names of the ATmega16 hardware registers and the actual hardware locations. When a register is used by name in the program, reference is made to the contents of that register.
- Function Prototypes
- Global Variables
- Main Program: We begin the main program by calling the function to initialize the ports and then enter a continuous loop. Within the loop body, the ATmega16 monitors for a status change on PORTB. When the user depresses one of the tact switches connected to PORTB, a change of status is detected and the appropriate LED is illuminated on PORTA.
- Function Definition

```
//***************************************************************
//file name: testbench.c
//function: provides test bench for ATMEL AVR ATmega16 controller
//target controller: ATMEL ATmega16
//
//ATMEL AVR ATmega16 Controller Pin Assignments
//Chip Port Function I/O Source/Dest Asserted Notes
//Pin 1 PB0 to active high RC debounced switch
//Pin 2 PB1 to active high RC debounced switch
//Pin 3 PB2 to active high RC debounced switch
```

```
//Pin 4 PB3 to active high RC debounced switch
//Pin 5 PB4 to active high RC debounced switch
//Pin 6 PB5 to active high RC debounced switch
//Pin 7 PB6 to active high RC debounced switch
//Pin 8 PB7 to active high RC debounced switch
//Pin 9 Reset
//Pin 10 VDD
//Pin 11 Gnd
//Pin 12 Resonator
//Pin 13 Resonator
//Pin 14 PD0 to tristate LED indicator
//Pin 15 PD1 to tristate LED indicator
//Pin 16 PD2 to tristate LED indicator
//Pin 17 PD3 to tristate LED indicator
//Pin 18 PD4 to tristate LED indicator
//Pin 19 PD5 to tristate LED indicator
//Pin 20 PD6 to tristate LED indicator
//Pin 21 PD7 to tristate LED indicator
//Pin 22 PC0
//Pin 23 PC1
//Pin 24 PC2
//Pin 25 PC3
//Pin 26 PC4
//Pin 27 PC5
//Pin 28 PC6
//Pin 29 PC7
//Pin 30 AVcc to VDD
//Pin 31 AGnd to Ground
//Pin 32 ARef to Vcc
//Pin 33 PA7
//Pin 34 PA6
//Pin 35 PA5
//Pin 36 PA4
//Pin 37 PA3
//Pin 38 PA2
//Pin 39 PA1
```

```c
//Pin 40 PA0
//
//author: Steven Barrett and Daniel Pack
//created: July 12, 2007
//last revised: July 12, 2007
//*************************************************************

//include files*************************************************
#include<iom16v.h>                    //ImageCraft ICC AVR
                                      //include file
//for ATmega16
//function prototypes*******************************************

void initialize_ports(void);          //initializes ports

//main program**************************************************
//global variables
unsigned char   old_PORTB = 0x00;    //present value of PORTB
unsigned char   new_PORTB;           //new values of PORTB
void main(void)
{
initialize_ports();                   //initialize ports
while(1){//main loop
  new_PORTB = PINB;                   //read PORTB

  if(new_PORTB != old_PORTB){         //process change
    //in PORTB input
    switch(new_PORTB){                //PORTB asserted high

            case 0x01:                //PB0 (0000_0001)
               PORTD=0x00;            //turn off all LEDs PORTD
               PORTD=0x01;            //turn on PD0 LED (0000_0001)
              break;

            case 0x02:                //PB1 (0000_0010)
               PORTD=0x00;            //turn off all LEDs PORTD
```

```c
            PORTD=0x02;             //turn on PD1 LED (0000_0010)
         break;

         case 0x04:                 //PB2 (0000_0100)
            PORTD=0x00;             //turn off all LEDs PORTD
            PORTD=0x04;             //turn on PD2 LED (0000_0100)
         break;

         case 0x08:                 //PB3 (0000_1000)
            PORTD=0x00;             //turn off all LEDs PORTD
            PORTD=0x08;             //turn on PD3 LED (0000_1000)
         break;

         case 0x10:                 //PB4 (0001_0000)
            PORTD=0x00;             //turn off all LEDs PORTD
            PORTD=0x10;             //turn on PD4 LED (0001_0000)
         break;

         case 0x20:                 //PB5 (0010_0000)
            PORTD=0x00;             //turn off all LEDs PORTD
            PORTD=0x20;             //turn on PD5 LED (0010_0000)
         break;

         case 0x40:                 //PB6 (0100_0000)
            PORTD=0x00;             //turn off all LEDs PORTD
            PORTD=0x40;             //turn on PD6 LED (0100_0000)
         break;

         case 0x80:                 //PB7 (1000_0000)
            PORTD=0x00;             //turn off all LEDs PORTD
            PORTD=0x80;             //turn on PD7 LED (1000_0000)
         break;

      default:;                     //all other cases
         }                          //end switch(new_PORTB)
      }                             //end if new_PORTB
```

```
        old_PORTB=new_PORTB;             //update PORTB
    }                                    //end while(1)
}                                        //end main
//*************************************************************
//function definitions
//*************************************************************


//*************************************************************
//initialize_ports: provides initial configuration for I/O ports
//*************************************************************

void initialize_ports(void)
{
DDRA=0xff;                               //set PORTA[7:0] as output
PORTA=0x00;                              //initialize PORTA[7:0] low

DDRB=0x00;                               //PORTB[7:0] as input
PORTB=0x00;                              //disable PORTB
    //pull-up resistors

DDRC=0xff;                               //set PORTC as output
PORTC=0x00;                              //initialize low

DDRD=0xff;                               //set PORTD as output
PORTD=0x00;                              //initialize low
}
//*************************************************************
```

1.7 PROGRAMMING THE ATmega16

Programming the ATmega16 requires several hardware and software tools. We briefly mention required components here. Please refer to the manufacturer's documentation for additional details.

Software tools: Throughout the text, we use the ImageCraft ICC AVR compiler. This is an excellent, user-friendly compiler. There are other excellent compilers available. The compiler is used to translate the source file (testbench.c) into machine language for loading into the ATmega16. We use Atmel's AVR Studio to load the machine code into the ATmega16.

Hardware tools: We use Atmel's STK500 AVR Flash MCU Starter Kit (STK500) for programming the ATmega16. The STK500 provides the interface hardware between the host PC and the ATmega16 for machine code loading. The STK500 is equipped with a complement of DIP sockets, which allows for programming all of the microcontrollers in the Atmel AVR line.

1.7.1 Programming Procedure

In this section, we provide a step-by-step procedure to program the ATmega16 DIP using the STK500 AVR Flash MCU Starter Kit. Please refer to Figure 1.5. It should be noted that the STK500 will be used to program the microcontroller, which will then be placed in the Testbench circuit.

1. Load AVR Studio (free download from www.atmel.com)
2. Ensure that the STK500 is powered down.
3. Connect the STK500 as shown in Figure 1.5.
4. Insert the ATmega16 into the red 40-pin socket. Note the location of pin 1 in Figure 1.5.
5. Power up the STK500.
6. Start up AVR Studio on your PC.
7. Pop-up window "Welcome to AVR Studio" should appear. Close this window by clicking on the "Cancel button."
8. Click on the "AVR icon." It looks like the silhouette of an integrated circuit. It is on the second line of the toolbar about halfway across the screen.
9. This should bring up a STK500 pop-up window with six tabs (Program, Fuses, Lockbits, Advanced, Board, and Auto). At the bottom of the Program tab window, verify that the STK500 was autodetected. Troubleshoot as necessary to ensure STK500 was autodetected by AVR Studio.
10. Set all tab settings:
    - Program:
        - Select device: ATmega16
        - Programming mode: ISP, Erase Device Before Programming, Verify Device After Programming
        - Flash: Input HEX file, Browse and find machine code file: yourfilename.hex
        - EEPROM: Input HEX file, Browse and find machine code file: yourfilename.EEP
    - Fuses: Set the following fuses
        - Boot flash section size = 128

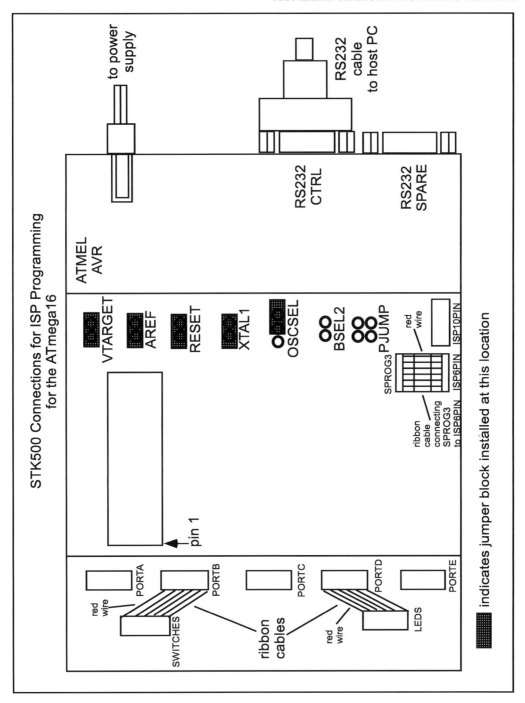

FIGURE 1.5: Programming the ATmega16 with the STK500.

- Brown out detection at Vcc = 4.0 V
- External Crystal/Resonator High Frequency; start-up time 16K CK + 64 ms
- Lock bits:
  - Mode 1
  - Application Protection Mode 1
  - Boot Loader Protection Mode 1
  - Advanced: N/A
  - Board: N/A
- Auto:
  - Erase Device
  - Program Flash
  - Verify Flash
  - Program Fuses
  - Verify Fuses
  - Read Fuses
11. Programming step:
    - Program Tab: click program
    - Fuse Tab: click program (do not forget this step---otherwise, processor runs very slow!)
12. Power down the STK500. Remove the programmed chip from the STK500 board and place it in the Testbench circuit.

1.8 SOFTWARE PORTABILITY

The software techniques discussed in the textbook are based on the ATmega16; however, the developed software may be easily ported for use with other Atmel AVR microcontrollers. To ease the transition to another microcontroller, it is suggested using a direct bit assignment technique. Instead of setting an entire register content at once, selected bits may be set. The individual bit definitions for the ATmega16 are provided in Appendix B.

For example, to set the UCSRB register, the following individual bit assignments may be used:

```
UCSRB = (1<<TXEN)|(1<<RXEN);        //Enable transmit and receive
```

as opposed to:

```
UCSRB = 0x08;                       //Enable transmit and receive
```

When transporting code, the header file for the specific microcontroller must be used, and also, the interrupt vector numbers may require change.

1.9 SUMMARY

In this chapter, we provided a brief overview of the ATmega16 microcontroller, a representative sample of the AVR microcontrollers. Information presented in this chapter can be readily applied to other microcontrollers in the AVR line. We then provided the Testbench hardware and software that we use throughout the text to illustrate peripheral subsystem operation aboard the ATmega16. In upcoming chapters, we provide additional details on selected ATmega16 subsystems.

1.10 REFERENCES AND FURTHER READING

1. S Barrett and D Pack, *Microcontroller Fundamentals for Engineers and Scientists*, Morgan & Claypool, San Rafael, CA, 2006. doi:10.2200/S00025ED1V01Y200605DCS001
2. Atmel 8-bit AVR Microcontroller with 16K Bytes In-System Programmable Flash, ATmega16, ATmega16L, data sheet: 2466L-AVR-06/05, Atmel, San Jose, CA.
3. J Hennessy and D Patterson, *Computer Architecture: A Quantitative Approach*, 3rd ed, Morgan Kaufman, San Francisco, CA, 2003.

1.11 CHAPTER PROBLEMS

1. Question: What is a RISC processor?
2. Question: How does the ATmega16 interact with the external world?
3. Question: What are the different methods of applying a clock source to the ATmega16? List the inherent advantages of each type.
4. Question: Describe the three different types of memory components aboard the ATmega16. What is each used for?
5. Question: Describe the three registers associated with each port.
6. Question: With a specific port, can some port pins be declared as output pins while others as input pins?
7. Question: Describe the serial communication features aboard the ATmega16. Provide a suggested application for each.
8. Question: What is the purpose of the ADC system?
9. Question: What is the purpose of the interrupt system?
10. Question: What is the purpose of the PWM system?
11. Question: What is the best clock speed to operate the ATmega16 at for a specific application?
12. Question: Sketch a flow chart or UML activity diagram for the testbench.c program.

CHAPTER 2

Serial Communication Subsystem

Objectives: After reading this chapter, the reader should be able to

- describe the differences between serial and parallel communication,
- provide definitions for key serial communications terminology,
- describe the operation of the USART,
- program the USART for basic transmission and reception,
- describe the operation of the SPI,
- program the SPI for basic transmission and reception, and
- describe the purpose of the two-wire interface (TWI).

2.1 SERIAL COMMUNICATIONS

Microcontrollers must often exchange data with other microcontrollers or peripheral devices. Data may be exchanged by using parallel or serial techniques. With parallel techniques, an entire byte of data is typically sent simultaneously from the transmitting device to the receiver device. Although this is efficient from a time point of view, it requires eight separate lines for the data transfer [1].

In serial transmission, a byte of data is sent a single bit at a time. Once 8 bits have been received at the receiver, the data byte is reconstructed. Although this is inefficient from a time point of view, it only requires a line (or two) to transmit the data.

The ATmega16 is equipped with a host of different serial communication subsystems, including the serial USART, SPI, and TWI. What all of these systems have in common is the serial transmission of data. Before discussing the different serial communication features aboard the ATmega16, we review serial communication terminology.

2.2 SERIAL COMMUNICATION TERMINOLOGY

In this section, we review common terminology associated with serial communication.

2.2.1 Asynchronous versus Synchronous Serial Transmission

In serial communications, the transmitting and receiving device must be synchronized to one another and use a common data rate and protocol. Synchronization allows both the transmitter and receiver to be expecting data transmission/reception at the same time. There are two basic methods of maintaining "sync" between the transmitter and receiver: asynchronous and synchronous.

In an asynchronous serial communication system, such as the USART aboard the ATmega16, framing bits are used at the beginning and end of a data byte. These framing bits alert the receiver that an incoming data byte has arrived and also signals the completion of the data byte reception. The data rate for an asynchronous serial system is typically much slower than the synchronous system, but it only requires a single wire between the transmitter and receiver.

A synchronous serial communication system maintains "sync" between the transmitter and receiver by employing a common clock between the two devices. Data bits are sent and received on the edge of the clock. This allows data transfer rates higher than with asynchronous techniques but requires two lines, data and clock, to connect the receiver and transmitter.

2.2.2 Baud Rate

Data transmission rates are typically specified as a baud or bits per second rate. For example, 9600 baud indicates data are being transferred at 9600 bits per second.

2.2.3 Full Duplex

Often, serial communication systems must both transmit and receive data. To do both transmission and reception simultaneously requires separate hardware for transmission and reception. A single duplex system has a single complement of hardware that must be switched from transmission to reception configuration. A full duplex serial communication system has separate hardware for transmission and reception.

2.2.4 Nonreturn to Zero Coding Format

There are many different coding standards used within serial communications. The important point is the transmitter and receiver must use a common coding standard so data may be interpreted correctly at the receiving end. The Atmel ATmega16 [2] uses a nonreturn to zero coding standard. In nonreturn to zero, coding a logic 1 is signaled by a logic high during the entire time slot allocated for a single bit, whereas a logic 0 is signaled by a logic low during the entire time slot allocated for a single bit.

2.2.5 The RS-232 Communication Protocol

When serial transmission occurs over a long distance, additional techniques may be used to ensure data integrity. Over long distances, logic levels degrade and may be corrupted by noise. At the receiving end, it is difficult to discern a logic high from a logic low. The RS-232 standard has been around for some time. With the RS-232 standard (EIA-232), a logic 1 is represented with a −12-VDC level, whereas a logic 0 is represented by a +12-VDC level. Chips are commonly available (e.g., MAX232) that convert the 5- and 0-V output levels from a transmitter to RS-232-compatible levels and convert back to 5- and 0-V levels at the receiver. The RS-232 standard also specifies other features for this communication protocol.

2.2.6 Parity

To further enhance data integrity during transmission, parity techniques may be used. Parity is an additional bit (or bits) that may be transmitted with the data byte. The ATmega16 uses a single parity bit. With a single parity bit, a single-bit error may be detected. Parity may be even or odd. In even parity, the parity bit is set to 1 or 0, such that the number of 1's in the data byte including the parity bit is even. In odd parity, the parity bit is set to 1 or 0, such that the number of 1's in the data byte including the parity bit is odd. At the receiver, the number of bits within a data byte including the parity bit are counted to ensure that parity has not changed, indicating an error, during transmission.

2.2.7 American Standard Code for Information Interchange

The American Standard Code for Information Interchange (ASCII) is a standardized seven-bit method of encoding alphanumeric data. It has been in use for many decades, so some of the characters and actions listed in the ASCII table are not in common use today. However, ASCII is still the most common method of encoding alphanumeric data. The ASCII code is provided in Figure 2.1. For example, the capital letter "G" is encoded in ASCII as 0x47. The "0x" symbol indicates the hexadecimal number representation. Unicode is the international counterpart of ASCII. It provides standardized 16-bit encoding format for the written languages of the world. ASCII is a subset of Unicode. The interested reader is referred to the Unicode home page website at www.unicode.org for additional information on this standardized encoding format.

2.3 SERIAL USART

The serial USART provide for full duplex (two-way) communication between a receiver and transmitter. This is accomplished by equipping the ATmega16 with independent hardware for the transmitter and receiver. The USART is typically used for asynchronous communication. That is, there is not a common clock between the transmitter and receiver to keep them synchronized with

		Most significant digit							
		0x0_	0x1_	0x2_	0x3_	0x4_	0x5_	0x6_	0x7_
Least significant digit	0x_0	NUL	DLE	SP	0	@	P	`	p
	0x_1	SOH	DC1	!	1	A	Q	a	q
	0x_2	STX	DC2	"	2	B	R	b	r
	0x_3	ETX	DC3	#	3	C	S	c	s
	0x_4	EOT	DC4	$	4	D	T	d	t
	0x_5	ENQ	NAK	%	5	E	U	e	u
	0x_6	ACK	SYN	&	6	F	V	f	v
	0x_7	BEL	ETB	'	7	G	W	g	w
	0x_8	BS	CAN	(	8	H	X	h	x
	0x_9	HT	EM	)	9	I	Y	i	y
	0x_A	LF	SUB	*	:	J	Z	j	z
	0x_B	VT	ESC	+	;	K	[	k	{
	0x_C	FF	FS	'	<	L	\	l	|
	0x_D	CR	GS	-	=	M	]	m	}
	0x_E	SO	RS	.	>	N	^	n	~
	0x_F	SI	US	/	?	O	_	o	DEL

FIGURE 2.1: ASCII Code. The ASCII code is used to encode alphanumeric characters. The "0x" indicates hexadecimal notation in the C programming language.

one another. To maintain synchronization between the transmitter and receiver, framing start and stop bits are used at the beginning and end of each data byte in a transmission sequence. The Atmel USART also has synchronous features. Space does not permit a discussion of these USART enhancements.

The ATmega16 USART is quite flexible. It has the capability to be set to a variety of data transmission or baud (bits per second) rates. The USART may also be set for data bit widths of 5 to 9 bits with one or two stop bits. Furthermore, the ATmega16 is equipped with a hardware-generated parity bit (even or odd) and parity check hardware at the receiver. A single parity bit allows for the detection of a single bit error within a byte of data. The USART may also be configured to operate in a synchronous mode. We now discuss the operation, programming, and application of the USART. Because of space limitations, we cover only the most basic capability of this flexible and powerful serial communication system.

2.3.1 System Overview

The block diagram for the USART is provided in Figure 2.2. The block diagram may appear a bit overwhelming, but realize there are four basic pieces to the diagram: the clock generator, the

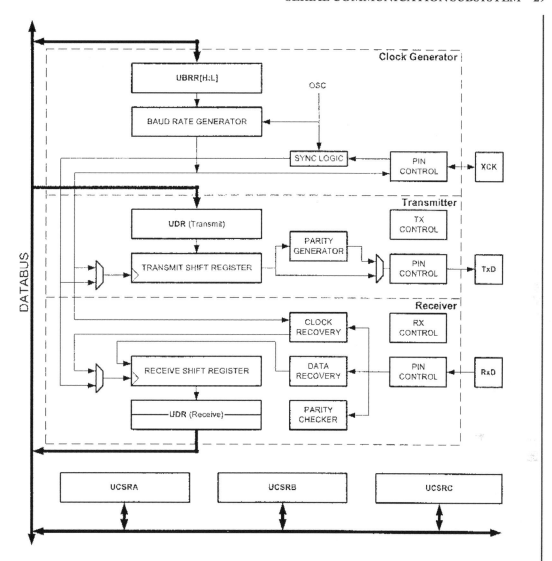

FIGURE 2.2: Atmel AVR ATmega16 USART block diagram. Figure used with permission of Atmel.

transmission hardware, the receiver hardware, and three control registers (UCSRA, UCSBR, and UCSRC). We discuss each in turn.

2.3.1.1 USART Clock Generator. The USART Clock Generator provides the clock source for the USART system and sets the baud rate for the USART. The baud rate is derived from the overall microcontroller clock source. The overall system clock is divided by the USART baud rate registers UBRR[H:L] and several additional dividers to set the baud rate. For the asynchronous normal mode (U2X bit = 0), the baud rate is determined using the following expression:

$$\text{baud rate} = (\text{system clock frequency})/(2(\text{UBRR} + 1)),$$

where UBRR is the content of the UBRRH and UBRRL registers (0–4095). Solving for UBRR yields

$$\text{UBRR} = ((\text{system clock generator})/(16 \times \text{baud rate})) - 1$$

2.3.1.2 USART Transmitter. The USART transmitter consists of a Transmit Shift Register. The data to be transmitted are loaded into the Transmit Shift Register via the USART I/O Data Register (UDR). The start and stop framing bits are automatically appended to the data within the Transmit Shift Register. The parity is automatically calculated and appended to the Transmit Shift Register. Data are then shifted out of the Transmit Shift Register via the TxD pin a single bit at a time at the established baud rate. The USART transmitter is equipped with two status flags: the USART Data Register Empty (UDRE) and the transmit complete (TXC) flags. The UDRE flag sets when the transmit buffer is empty, indicating it is ready to receive new data. This bit should be written to a zero when writing the USART Control and Status Register A (UCSRA). The UDRE bit is cleared by writing to the UDR. The TXC flag bit is set to logic 1 when the entire frame in the Transmit Shift Register has been shifted out and there are no new data currently present in the transmit buffer. The TXC bit may be reset by writing a logic 1 to it.

2.3.1.3 USART Receiver. The USART Receiver is virtually identical to the USART Transmitter except for the direction of the data flow, which is reversed. Data are received a single bit at a time via the RxD pin at the established baud rate. The USART receiver is equipped with the receive complete (RXC) flag. The RXC flag is logic 1 when unread data exist in the receive buffer.

2.3.1.4 USART Registers. In this section, we discuss the register settings for controlling the USART system. We have already discussed the function of the UDR and the USART baud rate registers (UBRRH and UBRRL). **Note:** The USART Control and Status Register C (UCSRC) and the USART baud rate register high (UBRRH) are assigned to the same I/O location in the memory map (Figure 2.3). The URSEL bit (bit 7 of both registers) determines which register

USART Control and Status Register A (UCSRA)

RXC	TXC	UDRE	FE	DOR	PE	U2X	MPCM
7							0

USART Control and Status Register B (UCSRB)

RXCIE	TXCIE	UDRIE	RXEN	TXEN	UCSZ2	RXB8	TXB8
7							0

USART Control and Status Register C (UCSRC)

URSEL=1	UMSEL	UPM1	UPM0	USBS	UCSZ1	UCSZ0	UCPOL
7							0

USART Data Register - UDR

UDR(Read)

RXB7	RXB6	RXB5	RXB4	RXB3	RXB2	RXB1	RXB0

UDR(Write)

TXB7	TXB6	TXB5	TXB4	TXB3	TXB2	TXB1	TXB0
7							0

USART Baud Rate Registers - UBRRH and UBRRL

UBRRH

URSEL=0	---	---	---	UBRR11	UBRR10	UBRR9	UBRR8

UBRRL

UBRR7	UBRR6	UBRR5	UBRR4	UBRR3	UBRR2	UBRR1	UBRR0
7							0

FIGURE 2.3: USART registers.

is being accessed. The URSEL bit must be 1 when writing to the UCSRC register and 0 when writing to the UBRRH register.

UCSRA. This contains the RXC, TXC, and the UDRE bits. The function of these bits has already been discussed.

UCSRB. This contains the receiver and transmitter enable bits (RXEN and TXEN, respectively). These bits are the "on/off" switch for the receiver and transmitter, respectively. The UCSRB register also contains the UCSZ2 bit. The UCSZ2 bit in the UCSRB register and the UCSZ[1:0] bits contained in the UCSRC register together set the data character size.

UCSRC. This allows the user to customize the data features to the application at hand. It should be emphasized that both the transmitter and receiver be configured with the same data features for proper data transmission. The UCSRC contains the following bits:

- USART mode select (UMSEL): 0, asynchronous operation; 1, synchronous operation
- USART parity mode (UPM[1:0]): 00, no parity; 10, even parity; 11, odd parity
- USART stop bit select (USBS): 0, one stop bit; 1, two stop bits
- USART character size (data width) (UCSZ[2:0]): 000, 5 bits; 001, 6 bits; 010; 7 bits; 011, 8 bits; 111, 9 bits

2.3.2 System Operation and Programming

The basic activities of the USART system consist of initialization, transmission, and reception. These activities are summarized in Figure 2.4. Both the transmitter and receiver must be initialized with the same communication parameters for proper data transmission. The transmission and reception activities are similar except for the direction of data flow. In transmission, we monitor for the UDRE flag to set, indicating the data register is empty. We then load the data for transmission into the UDR register. For reception, we monitor for the RXC bit to set, indicating there are unread data in the UDR register. We then retrieve the data from the UDR register.

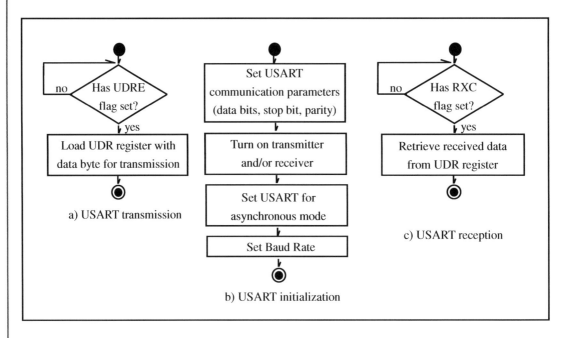

FIGURE 2.4: USART activities.

SERIAL COMMUNICATION SUBSYSTEM

To program the USART, we implement the flow diagrams provided in Figure 2.4. In the sample code provided, we assume the ATmega16 is operating at 10 MHz and we desire a baud rate of 9600, asynchronous operation, no parity, one stop bit, and eight data bits.

To achieve 9600 baud with an operating frequency of 10 MHz requires that we set the UBRR registers to 64, which is 0x40.

```
//**************************************************************
//USART_init: initializes the USART system
//**************************************************************

void USART_init(void)
{
UCSRA = 0x00;                //control register initialization
UCSRB = 0x08;                //enable transmitter
UCSRC = 0x86;                //async, no parity, 1 stop bit,
                             //8 data bits
                             //Baud Rate initialization

UBRRH = 0x00;
UBRRL = 0x40;
}



//**************************************************************
//USART_transmit: transmits single byte of data
//**************************************************************

void USART_transmit(unsigned char data)
{
while((UCSRA & 0x20)==0x00)  //wait for UDRE flag
   {
   ;
   }
UDR = data;                  //load data to UDR for transmission
}
```

```
//***************************************************************
//USART_receive: receives single byte of data
//***************************************************************

unsigned char USART_receive(void)
{
while((UCSRA & 0x80)==0x00)   //wait for RXC flag
   {
   ;
   }
data = UDR;                   //retrieve data from UDR
return data;
}


//***************************************************************
```

2.3.3 Serial Peripheral Interface

The ATmega16 SPI also provides for two-way serial communication between a transmitter and a receiver. In the SPI system, the transmitter and receiver share a common clock source. This requires an additional clock line between the transmitter and receiver but allows for higher data transmission rates as compared with the USART. The SPI system allows for fast and efficient data exchange between microcontrollers or peripheral devices. There are many SPI-compatible external systems available to extend the features of the microcontroller. For example, a liquid crystal display (LCD) or a digital-to-analog converter (DAC) could be added to the microcontroller using the SPI system.

2.3.3.1 SPI Operation. The SPI may be viewed as a synchronous 16-bit shift register with an 8-bit half residing in the transmitter and the other 8-bit half residing in the receiver as shown in Figure 2.5. The transmitter is designated the master because it provides the synchronizing clock source between the transmitter and the receiver. The receiver is designated as the slave. A slave is chosen for reception by taking its slave select ($\overline{SS}$) line low. When the $\overline{SS}$ line is taken low, the slave's shifting capability is enabled. SPI transmission is initiated by loading a data byte into the master configured SPI Data Register (SPDR). At that time, the SPI clock generator provides clock pulses to the master and also to the slave via the SCK pin. A single bit is shifted out of the master designated shift register on the Master Out Slave In (MOSI) microcontroller pin on every

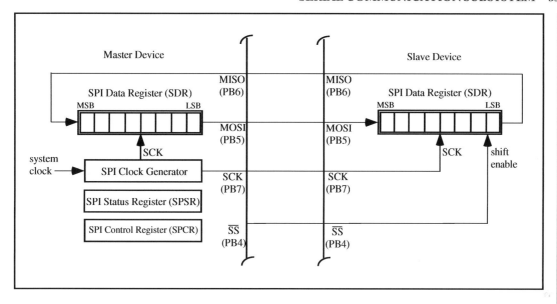

FIGURE 2.5: SPI overview.

SCK pulse. The data are received at the MOSI pin of the slave designated device. At the same time, a single bit is shifted out of the Master In Slave Out (MISO) pin of the slave device and into the MISO pin of the master device. After eight master SCK clock pulses, a byte of data has been exchanged between the master and slave designated SPI devices. Completion of data transmission in the master and data reception in the slave is signaled by the SPI Interrupt Flag (SPIF) in both devices. The SPIF flag is located in the SPI Status Register (SPSR) of each device. At that time, another data byte may be transmitted.

2.3.3.2 Registers. The registers for the SPI system are provided in Figure 2.6. We will discuss each one in turn.

SPI Control Register. The SPI Control Register (SPCR) contains the "on/off" switch for the SPI system. It also provides the flexibility for the SPI to be connected to a wide variety of devices with different data formats. It is important that both the SPI master and slave devices be configured for compatible data formats for proper data transmission. The SPCR contains the following bits:

- SPI Enable (SPE) is the "on/off" switch for the SPI system. A logic 1 turns the system on and logic 0 turns it off.

SPI Control Register - SPCR							
SPIE	SPE	DORD	MSTR	CPOL	CPHA	SPR1	SPR0
7							0

SPI Status Register - SPSR							
SPIF	WCOL	---	---	---	---	---	SPI2X
7							0

SPI Data Register - SPDR							
MSB							LSB
7							0

FIGURE 2.6: SPI registers.

- Data Order (DORD) allows the direction of shift from master to slave to be controlled. When the DORD bit is set to 1, the least significant bit (LSB) of the SPDR is transmitted first. When the DORD bit is set to 0, the Most Significant Bit (MSB) of the SPDR is transmitted first.
- The Master/Slave Select (MSTR) bit determines if the SPI system will serve as a master (logic 1) or slave (logic 0).
- The Clock Polarity (CPOL) bit allows determines the idle condition of the SCK pin. When CPOL is 1, SCK will idle logic high, whereas when CPOL is 0, SCK will idle logic 0.
- The Clock Phase (CPHA) determines if the data bit will be sampled on the leading (0) or trailing (1) edge of the SCK.
- The SPI SCK is derived from the microcontroller's system clock source. The system clock is divided down to form the SPI SCK. The SPI Clock Rate Select (SPR[1:0]) bits and the Double SPI Speed (SPI2X) bit are used to set the division factor. The following divisions may be selected using SPI2X, SPR1, and SPR0:
    - 000: SCK = system clock/4
    - 001: SCK = system clock/16
    - 010: SCK = system clock/64
    - 011: SCK = system clock/1284
    - 100: SCK = system clock/2
    - 101: SCK = system clock/8

- 110: SCK = system clock/32
- 111: SCK = system clock/64

SPI Status Register. This contains the SPIF. The flag sets when eight data bits have been transferred from the master to the slave. The SPIF bit is cleared by first reading the SPSR after the SPIF flag has been set and then reading the SPDR. The SPSR also contains the SPI2X bit used to set the SCK frequency.

SPI Data Register. As previously mentioned, writing a data byte to the SPDR initiates SPI transmission.

2.3.3.3 Programming. To program the SPI system, the system must first be initialized with the desired data format. Data transmission may then commence. Functions for initialization, transmission, and reception are provided below. In this specific example, we divide the clock oscillator frequency by 128 to set the SCK clock frequency.

```
//***********************************************************
//spi_init: initializes spi system
//***********************************************************

void spi_init(unsigned char control)
{
DDRB = 0xA0;            //Set SCK (PB7), MOSI (PB5) for output,
                            others to input
                        //Configure SPI Control Register (SPCR)
SPCR = 0x53;            //SPIE:0,SPE:1,DORD:0,MSTR:1,CPOL:0,CPHA:0,
                            SPR:1,SPR0:1
}


//***********************************************************
//spi_write: Used by SPI master to transmit a data byte
//***********************************************************

void spi_write(unsigned char byte)
{
```

```
SPDR = byte;
while (!(SPSR & 0x80));
}


//*************************************************************
//spi_read: Used by SPI slave to receive data byte
//*************************************************************

unsigned char spi_read(void)
{
while (!(SPSR & 0x80));

return SPDR;
}


//*************************************************************
```

2.4 TWO-WIRE SERIAL INTERFACE

The TWI subsystem allows the system designer to network a number of related devices (microcontrollers, transducers, displays, memory storage, etc.) together into a system using a two-wire interconnecting scheme. The TWI allows a maximum of 128 devices to be connected together. Each device has its own unique address and may both transmit and receive over the two-wire bus at frequencies up to 400 kHz. This allows the device to freely exchange information with other devices in the network within a small area. Space does not permit a detailed discussion of this advanced serial communication system.

2.5 SUMMARY

In this chapter, we have discussed the differences between parallel and serial communications and key serial communication-related terminology. We then, in turn, discussed the operation of USART, SPI, and TWI serial communication systems. We also provided basic code examples to communicate with the USART and SPI systems.

2.6 REFERENCES AND FURTHER READING

1. S Barrett and D Pack, *Microcontroller Fundamentals for Engineers and Scientists*, Morgan & Claypool, San Rafael, CA, 2006. doi:10.2200/S00025ED1V01Y200605DCS001

2. *Atmel 8-bit AVR Microcontroller with 16K Bytes In-System Programmable Flash, ATmega16, ATmega16L**, data sheet: 2466L-AVR-06/05, Atmel, San Jose, CA.

2.7 CHAPTER PROBLEMS

1. Question: Summarize the differences between parallel and serial conversion.
2. Question: Summarize the differences among the USART, SPI, and TWI methods of serial communication.
3. Question: Draw a block diagram of the USART system, label all key registers, and all keys USART flags.
4. Question: Draw a block diagram of the SPI system, label all key registers, and all keys USART flags.
5. Question: If an ATmega16 microcontroller is operating at 12 MHz, what is the maximum transmission rate for the USART and the SPI?
6. Question: What is the ASCII encoded value for "Claypool"?
7. Question: Draw the schematic of a system consisting of two ATmega16 that will exchange data via the SPI system. The system should include RS-232 level shifting.
8. Question: Write the code to implement the system described in the question above.
9. Question: Add USART and SPI features to the testbench.

CHAPTER 3
Analog-to-Digital Conversion

Objectives: After reading this chapter, the reader should be able to

- explain the difference between analog and digital signals,
- illustrate the ADC process,
- assess the quality of ADC using the metrics of sampling rate, quantization levels, number of bits used for encoding, and dynamic range,
- design signal conditioning circuits to interface sensors with ADCs,
- describe the key registers used during an ATmega16 ADC,
- describe the steps to perform an ADC with the ATmega16, and
- program the ATmega16 to perform an ADC.

A microcontroller is used to process information from the natural world, decide on a course of action based on the information collected, and then issue control signals to implement the decision. Because much of the information from the natural world is analog or continuous in nature and the microcontroller is a digital or discrete-based processor, a method to convert an analog signal to digital is required [1]. An ADC system performs this task, whereas a DAC performs the conversion in the opposite direction. We will discuss both types of converters in this chapter.

In the first section, we present the fundamental concepts associated with the ADC process. In the following section, we discuss the conversion process itself, followed by a presentation of different hardware implementations of the process. Much of these early sections contain the same material you will find in our text, *Microcontroller Fundamentals for Engineers and Scientists*. We then review the basic features of the ATmega16 ADC system, followed by a system description and a discussion of key ADC registers. We conclude our discussion of the ADC with several illustrative code examples. We conclude the chapter with a discussion of the DAC process.

3.1 BACKGROUND THEORY
Before we discuss the ADC process, we need to familiarize you with underlying theories that support the process. We start with some definitions on analog and digital signals.

3.1.1 Analog versus Digital Signals

A signal is a collection of values representing the state of a physical variable. The collection can be as small as only one value or can have as many values as you wish. In engineering, we usually arrange the values in order, for example, over time or over a spatial axis, to display the information. The time and spatial variables are called independent variables, because they are not affected by the physical variables of interests. For example, we measure the temperature change over time. The temperature measured is dependent on the time, not the other way around. Figure 3.1 shows an altitude trajectory of a bird flying over a short period. The signal shows how the altitude of the bird changes continuously.

Figure 3.2 shows a grayscale image of a six-legged robot. The image captured the light intensities of the scene using a charge-coupled device camera. If we move from the left to the right on the image and observe the intensity changes, we can find vertical edges by detecting signal

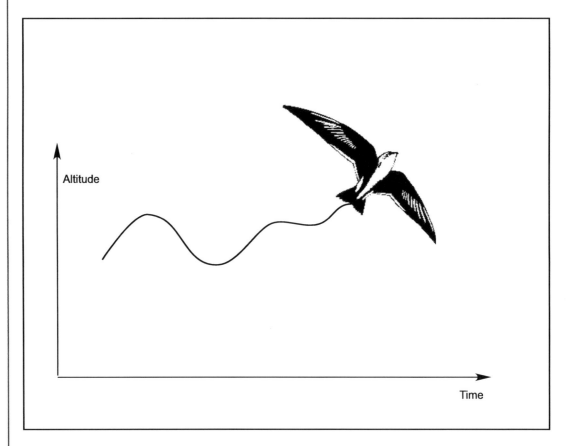

FIGURE 3.1: Altitude trajectory generated by a flying bird.

ANALOG-TO-DIGITAL CONVERSION 43

FIGURE 3.2: A photo of a walking robot.

intensity changes in the spatial axis. The same analysis can be performed as we move from the top to the bottom of the image.

Analog signals are those whose physical variable values change continuously over their independent variable. Most physical variables, your speech, movement of stars, and the music you hear at a concert are analog signals, signals that we find all around us. Digital signals, on the other hand, have their physical variables defined only for discrete instances over their independent variables. Although it may look continuous to human eyes, the photo example shown in Figure 3.2 is a discrete signal because pixels that make up a camera image cannot capture all space within the camera frame. The image is only a finite composition of intensity values seen by a discrete number of pixels.

Digital signals are important because all signals represented in digital systems, computers, and microcontrollers are in digital forms. The important task is how to faithfully represent analog signals using digital signals. For example, human voices must be converted to corresponding digital signals before they can be routed by digital switching circuits in telephone communication systems. Similarly, voice commands to robots must be converted to a digital form before robots can understand the command.

As shown in the examples above, we live in an analog world; that is, physical variables are analog signals. It is precisely this reason why the ADC is so very important in any digital systems that interact with an analog environment.

3.1.2 Sampling, Quantization, and Encoding

In this subsection, we present three important processes associated with the ADC. We first start with the subject of sampling. Imagine yourself as a photographer in an Olympic diving stadium. Your job is taking a sequence of pictures of divers jumping off from a diving board 10 meters above the surface of the diving pool. Your goal is to put the sequence of pictures together to reconstruct the motion of each diver. The sequence of pictures makes up samples of divers' motions. If a diver tries a complex motion and you want to faithfully reconstruct his motion, you must take enough pictures from the start to the end of the dive. If a diver makes a simple routine dive, you only need to take a few pictures over the period of the dive. Two very different cases of motions generated by a diver is shown in Figure 3.3. The same time sequence is used to capture samples for both motions. As can be seen from figure, frame (a) motion cannot be regenerated from the samples, whereas the motion shown in frame (b) can clearly be reconstructed from the same number of samples used to capture both motions.

Sampling is the process of taking "snapshots" of a signal over time. Naturally, when we sample a signal, we want to sample it in an optimal fashion such that we can capture the essence of the signal while minimizing the use of resources. In essence, we want to minimize the number of samples while faithfully reconstructing the original signal from the samples. As can be deduced from our discussion above, the rate of change in a signal determines the number of samples required to faithfully reconstruct the signal, provided that all adjacent samples are captured with the same sample timing intervals.

Harry Nyquist from Bell Laboratory studied the sampling process and derived a criterion that determines the minimum sampling rate for any continuous analog signals. His, now famous, minimum sampling rate is known as the Nyquist sampling rate, which states that one must sample a signal at least twice as fast as the highest frequency content of the signal of interest. For example, if we are dealing with the human voice signal that contains frequency components that span from about 20 Hz to 4 kHz, the Nyquist sample theorem tells us that we must sample the signal at least at 8 kHz, 8000 "snapshots" every second. Engineers who work for telephone companies must deal with such issues. For further study on the Nyquist sampling rate, refer to Barrett

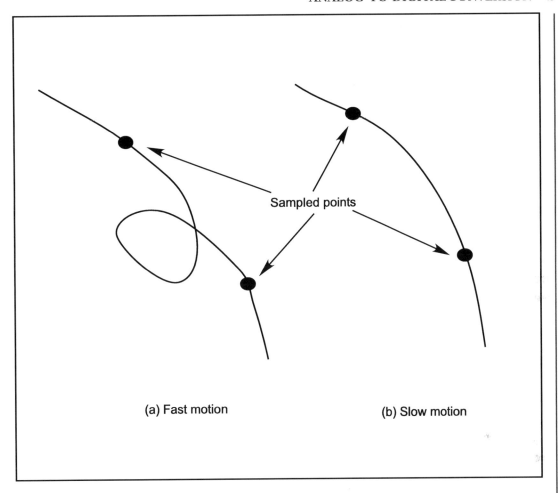

FIGURE 3.3: Two divers jumping off the platforms: (a) fast motion and (b) slow motion.

and Pack [2] listed in the References section. Sampling is important because when we want to represent an analog signal in a digital system, such as a computer, we must use the appropriate sampling rate to capture the analog signal for a faithful representation in digital systems.

Now that we understand the sampling process, let us move on to the second process of the ADC, quantization. Each digital system has a number of bits, which it uses as the basic units to represent data. A bit is the most basic unit where single binary information, 1 or 0, is represented. A nibble is made up of 4 bits put together. A byte is 8 bits.

In the previous section, we tacitly avoided the discussion of the form of captured signal samples. When a signal is sampled, digital systems need some means to represent the captured samples. The quantization of a sampled signal is how the signal is represented as one of quantization level. Suppose you have a single bit to represent an incoming signal. You only have two different numbers, 0 and 1. You may say that you can distinguish only low from high. Suppose you have 2 bits. You can represent four different levels, 00, 01, 10, and 11. What if you have 3 bits? You now can represent eight different levels: 000, 001, 010, 011, 100, 101, 110, and 111. Think of it as follows. When you had 2 bits, you were able to represent four different levels. If we add one more bit, that bit can be 1 or 0, making the total possibilities 8. Similar discussion can lead us to conclude that given n bits, we have 2^n different numbers or levels one can represent.

Figure 3.4 shows how n bits are used to quantize a range of values. In many digital systems, the incoming signals are voltage signals. The voltage signals are first obtained from physical signals with the help of transducers, such as microphones, angle sensors, and infrared sensors. The voltage signals are then conditioned to map their range with the input range of a digital system, typically 0 to 5 V. In Figure 3.4, n bits allow you to divide the input signal range of a digital system into

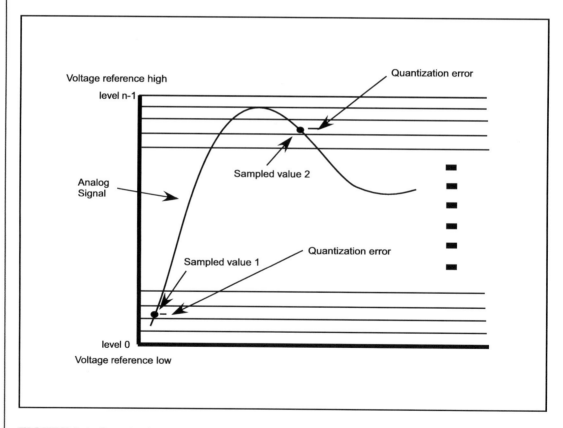

FIGURE 3.4: Quantization.

ANALOG-TO-DIGITAL CONVERSION 47

2^n different quantization levels. As can be seen from the figure, higher quantization levels means better mapping of an incoming signal to its true value. If we only had a single bit, we can only represent levels 0 and 1. Any analog signal value in between the range had to be mapped either as level 0 or level 1, not many choices. Now imagine what happens as we increase the number of bits available for the quantization levels. What happens when the available number of bits is 8? How many different quantization levels are available now? Yes, 256. How about 10, 12, or 14? Notice also that as the number of bits used for the quantization levels increases for a given input range the 'distance' between two adjacent levels decreases with a factor of a polynomial.

Finally, the encoding process involves converting a quantized signal into a digital binary number. Suppose again we are using 8 bits to quantize a sampled analog signal. The quantization levels are determined by the 8 bits, and each sampled signal is quantized as one of 256 quantization levels. Consider the two sampled signals shown in Figure 3.5. The first sample is mapped to quantization level 2 and the second one is mapped to quantization level 198. Note the amount

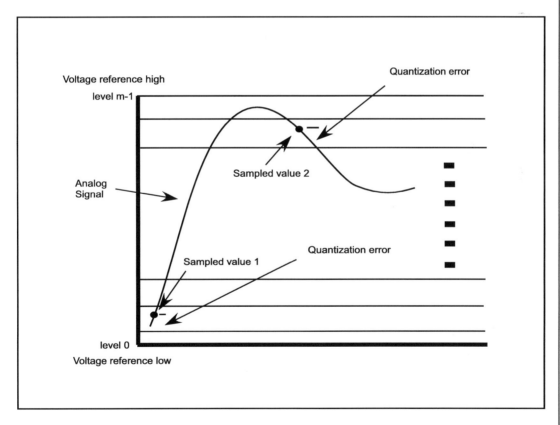

FIGURE 3.5: Quantization with fewer bits.

of quantization error introduced for both samples. Now consider Figure 3.5. The same signal is sampled at the same time but quantized using a less number of bits. Note that the quantization error is inversely proportional to the number of bits used to quantize the signal.

Once a sampled signal is quantized, the encoding process involves representing the quantization level with the available bits. Thus, for the first sample, the encoded sampled value is 0000 0001, whereas the encoded sampled value for the second sample is 1100 0110. As a result of the encoding process, sampled analog signals are now represented as a set of binary numbers. Thus, the encoding is the last necessary step to represent a sampled analog signal into its corresponding digital form, shown in Figure 3.6.

3.1.3 Resolution and Data Rate

Resolution is a measure used to quantize an analog signal. In fact, resolution is nothing but the "distance" between two adjacent quantization levels we discussed earlier. Suppose again we have

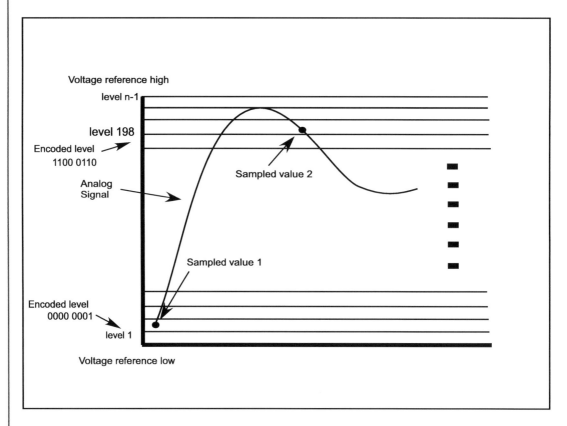

FIGURE 3.6: Encoding.

a range of 5 V and 1 bit to represent an analog signal. The resolution in this case is 2.5 V, a very poor resolution. You can imagine how your TV screen will look if you only had only two levels to represent each pixel, black and white. The maximum error, called the resolution error, is 2.5 V for the current case, 50% of the total range of the input signal. Suppose you now have 4 bits to represent quantization levels. The resolution now becomes 1.25 V, or 25% of the input range. Suppose you have 20 bits for quantization levels. The resolution now becomes 4.77×10^{-6}, 9.54×10^{-5}% of the total range. The discussion we presented simply illustrates that as we increase the available number of quantization levels within a range, the distance between adjacent levels decreases, reducing the quantization error of a sampled signal. As the number grows, the error decreases, making the representation of a sampled analog signal more accurate in the corresponding digital form. The number of bits used for the quantization is directly proportional to the resolution of a system. You now should understand the technical background when you watch high-definition television broadcasting.

Now let us move onto the discussion of the data rate. The definition of the data rate is the amount of data generated by a system per some time unit. Typically, the number of bits or the number of bytes per second is used as the data rate of a system. We just saw that the more bits we use for the quantization levels, the more accurate we can represent a sampled analog signal. Why not use the maximum number of bits current technologies can offer for all digital systems, when we convert analog signals to digital counterparts? It has to do with the cost involved. In particular, suppose you are working for a telephone company and your switching system must accommodate 100,000 customers. For each individual phone conversation, suppose the company uses an 8-kHz sampling rate and you are using 10 bits for the quantization levels for each sampled signal.[1] If all customers are making out-of-town calls, what is the number of bits your switching system must process to accommodate all calls? The answer will be $100,000 \times 8000 \times 10$, or 8 billion bits per every second! You will need some major computing power to meet the requirement. For such reasons, when designers make decisions on the number of bits used for the quantization levels, they must consider the computational burden the selection will produce on the computational capabilities of a digital system versus the required system resolution.

You will also encounter the term *dynamic range* when you consider finding appropriate ADCs. The dynamic range is a measure used to describe the signal to noise ratio. The unit used for the measurement is decibel, which is the strength of a signal with respect to a reference signal. The greater the decibel number, the stronger the signal is compared with a noise signal. The definition of the dynamic range is $20 \log 2^b$, where b is the number of bits used to convert analog signals

---

[1] For the sake of our discussion, we ignore other overheads involved in processing a phone call such as multiplexing, demultiplexing, and serial-to-parallel conversion.

to digital signals. Typically, you will find 8 to 12 bits used in commercial ADCs, translating the dynamic range from 20 log 2^8 dB to 20 log 2^{12} dB (Oppenheim and Schafer [3]).

3.2 ANALOG-TO-DIGITAL CONVERSION PROCESS

The goal of the ADC process is to accurately represent analog signals as digital signals. Toward this end, three signal processing procedures, sampling, quantization, and encoding, described in the previous section must be combined together. Before the ADC process takes place, we first need to convert a physical signal into an electrical signal with the help of a transducer. A transducer is an electrical and/or mechanical system that converts physical signals into electrical signals or electrical signals to physical signals. Depending on the purpose, we categorize a transducer as an input transducer or an output transducer. If the conversion is from physical to electrical, we call it an input transducer. The mouse, the keyboard, and the microphone for your PC all fall under this category. A camera, an infrared sensor, and a temperature sensor are also input transducers. The output transducer converts electrical signals to physical signals. The computer screen and the printer for your computer are output transducers. Speakers and electrical motors are also output transducers. Therefore, transducers play the central part for digital systems to operate in our physical world by transforming physical signals to and from electrical signals.

In addition to transducers, we also need signal conditioning circuitry before we apply the ADC process or its opposite process of DAC process. The signal conditioning circuitry is called the transducer interface. The objective of the transducer interface circuit is to scale and shift the electrical signal range to map the output of the input transducer to the input of the ADC. Figure 3.7 shows the transducer interface circuit using an input transducer.

The output of the input transducer is first scaled by constant K. In the figure, we use a microphone as the input transducer whose output ranges from -5VDC to $+5$ VDC. The input to the ADC ranges from 0 to 5 VDC. The box with constant K maps the output range of the input transducer to the input range of the converter. Naturally, we need to multiply all input signals by $1/2$ to accommodate the mapping. Once the range has been mapped, the signal now needs to be shifted. Note that the scale factor maps the output range of the input transducer as -2.5 to $+2.5$ VDC instead of 0 to 5 VDC. The second portion of the circuit shifts the range by 2.5 VDC, thereby completing the correct mapping. Actual implementation of the circuit components is accomplished using amplifiers with some feedback loops.

In general, the scaling and bias process may be described by two equations:

$$V_{2\,max} = (V_{1\,max} \times K) + B$$

$$V_{2\,min} = (V_{1\,min} \times K) + B.$$

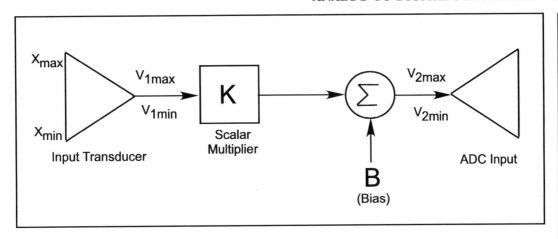

FIGURE 3.7: A block diagram of the signal conditioning for an ADC. The range of the sensor voltage output is mapped to the ADC input voltage range. The scalar multiplier maps the magnitudes of the two ranges, and the bias voltage is used to align two limits.

The variable $V_{1\,max}$ represents the maximum output voltage from the input transducer. This voltage occurs when the maximum physical variable (X_{max}) is presented to the input transducer. This voltage must be scaled by the scalar multiplier (K) and then have a DC offset bias voltage (B) added to provide the voltage $V_{2\,max}$ to the input of the ADC converter.

Similarly, The variable $V_{1\,min}$ represents the minimum output voltage from the input transducer. This voltage occurs when the minimum physical variable (X_{min}) is presented to the input transducer. This voltage must be scaled by the scalar multiplier (K) and then have a DC offset bias voltage (B) added to produce voltage $V_{2\,min}$ to the input of the ADC converter.

Usually, the values of $V_{1\,max}$ and $V_{1\,min}$ are provided with the documentation for the transducer. Also, the values of $V_{2\,max}$ and $V_{2\,min}$ are known. They are the high and low reference voltages for the ADC system (usually 5 and 0 VDC for a microcontroller). We thus have two equations and two unknowns to solve for K and B. The circuits to scale by K and add the offset B are usually implemented with operational amplifiers. We refer interested readers to Thomas and Rosa [4] listed in the References section.

Once a physical signal has been converted to its corresponding electrical signal with the help of an input transducer and the output of the transducer mapped correctly to the input of the ADC, the ADC process can start. The first step of the ADC process is the sampling of the analog signal. When selecting a converter, one must consider the type of physical signal that is being converted to properly ensure the sampling rate. As discussed in the previous section, using the proper sampling rate is the first step that determines whether an analog signal will be represented correctly in digital systems. What this means for the reader is to select an ADC that can handle a required conversion rate. Because most microcontrollers now come with a built-in ADC, one must study the user manual portion discussing the conversion rate and make sure that the required sampling rate for the application falls under the advertised conversion rate. For example, if you need to convert the signal representing a person's blood pressure, the sampling rate with 100 Hz (100 samples per second) will suffice. On the other hand, if you are dealing with human voice, you need at least an 8-kHz sampling rate capacity (see Enderle et al. [5] for details).

Once the analog signal has been sampled, the quantization process takes place. For this process, again one must decide how much quantization error can be allowed. At one extreme where you are only concerned with finding out only two states, say on and off, quantization error of 1 V is not important. We can operate safely with 2 bits with maximum quantization error of 1.25 V. On the other hand, if we can only operate with maximum quantization error of 0.01 V, we need to choose a converter with, at minimum, 10 bits (about 5 mV). To determine the number of bits and its corresponding maximum quantization error, we use the following equation.

$$\text{Resolution} = \frac{\text{range}}{2^b}.$$

Thus, we can determine the number of bits that will meet the error requirement using the equation above.

Once the quantization level has been determined, we can now encode it using the available bits. As seen in the previous section, the process is a simple conversion of a decimal number (quantization level) to a binary number. Note that the binary number should use all available bits. For example, quantization level 4 using 8 bits is converted as 0000 0100, not 100.

In summary, the ADC process has three steps for completion: sampling, quantization, and encoding. In the next section, we delve into four different technologies used to implement the ADC process.

Example 3.1 A photodiode is a semiconductor device that provides an output current corresponding to the light impinging on its active surface. The photodiode is used with a transimpedance amplifier to convert the output current to an output voltage. A photodiode/transimpedance amplifier provides an output voltage of 0 V for maximum rated light intensity and −2.50 VDC of output voltage for the minimum rated light intensity. Calculate the required values of K and B for this light transducer, so it may be interfaced to a microcontroller's ADC system.

$$V_{2\ max} = (V_{1\ max} \times K) + B$$

$$V_{2\ min} = (V_{1\ min} \times K) + B$$

$$5.0 \text{ V} = (0 \text{ V} \times K) + B$$

$$0 \text{ V} = (-2.50 \text{ V} \times K) + B$$

The values of K and B may then be determined to be 2 and 5 VDC, respectively.

3.3 ADC CONVERSION TECHNOLOGIES

The ATmega16 uses a successive approximation converter technique to convert an analog sample into a 10-bit digital representation. In this section, we will discuss this type of conversion process as well as other techniques. In certain applications, you are required to use converter technologies external to the microcontroller.

We begin by briefly discussing four different types of technologies used in ADCs: the successive approximation conversion, the integration conversion, the counter-based conversion, and the parallel conversion. For a detailed discussion, we refer the interested readers to Barrett and Pack [2]. For the following discussion of different ADC technologies, see Figure 3.8. One can group all existing technologies into two different categories: direct conversion versus indirect conversion. The successive approximation conversion, counter-based conversion, and parallel conversion use techniques to directly convert analog input signals into their digital counterparts. They are all considered to fall under the direct conversion category. The integration method uses conversion time as the means to indirectly measure the magnitude of the incoming analog signals, an indirect approach.

3.3.1 Successive Approximation

The successive approximation technique uses a DAC, a controller, and a comparator to perform the ADC process. Starting from the MSB down to the LSB, the controller turns on each bit at a time and generates an analog signal, with the help of the DAC, to be compared with the original input analog signal. Based on the result of the comparison, the controller changes or leaves the current bit and turns on the next MSB. The process continues until decisions are made for all available bits. One can consider the process similar to a game children play often. One child thinks of a number in the range of 0 to 10 and asks another child to guess the number within n turns. The first child will tell the second child whether a guessed number is higher or lower than the answer at the end of each turn. The optimal strategy in such a situation is to guess the middle number in the range, say 5. If the answer is higher than 5, the second guess should be 8. If the answer is lower than 5, the second guess should be 3. The strategy is to narrow down to the answer by partitioning the available range into two equal parts at every turn. The successive approximation method works similarly in that the MSB is used to partition the original input range of an ADC into halves, the

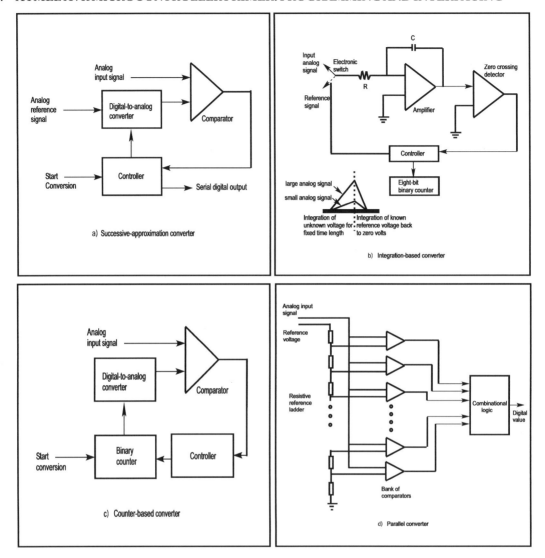

FIGURE 3.8: Four different technologies used to convert analog signals to digital signals: (a) successive approximation converter, (b) integration-based converter, (c) counter-based converter, and (d) parallel ADC.

second MSB divides the remaining half into two quarters of the input range, and so forth. Figure 3.8(a) shows the architecture of this type of converter. The advantage of this technique is that the conversion time is uniform for any input, but the disadvantage of the technology is the use of complex hardware for implementation.

3.3.2 Integration

The integration technique uses an integrator, a comparator, and a controller to convert analog signals to digital signals. A sampled analog signal is integrated over a fixed period, say n clock cycles of the digital system. Another fixed reference signal is integrated over time and compared with the input analog signal integrated. Although the value of the reference signal integrated is smaller than the input analog signal integrated, the reference signal is continuously integrated, and the time for the integration is measured. When the two integrated values equal, the measured time is converted to a digital encoded value. Figure 3.8(b) shows the system components of the converter. One disadvantage of this technique is the varying time for the conversion process. A small analog value will take less time to convert compared with a large value.

3.3.3 Counter-Based Conversion

The third technique to convert an analog signal to a digital signal is the counter-based conversion. This conversion is performed with the help of a counter, a DAC, and a comparator. The counter starts at 0 and counts up. As the counter counts up, the corresponding value is converted to an analog value and compared with an input analog signal. As long as the input analog signal is greater than the signal generated by the DAC, the counter counts up and the process continues. When the comparator detects that the signal from the DAC is greater than the input analog signal, the counter value is then converted to a digital value representing the sampled analog signal. Figure 3.8(c) shows the overall architecture of this converter. Similar to the converter based on the integration principle, the disadvantage of this type of converter is the varying conversion time.

3.3.4 Parallel Conversion

The last technique allows the quickest conversion time among the techniques we discussed. A parallel converter uses a large number of comparators and circuitry to simultaneously measure the input signal and convert it to a digital value. The obvious disadvantage of this technique is the cost involved in building the circuitry. Figure 3.8(d) shows the architecture of the converter.

3.4 THE ATMEL ATmega16 ADC SYSTEM

The Atmel ATmega16 [6] microcontroller is equipped with a flexible and powerful ADC system. It has the following features:

- 10-bit resolution,
- ± 2 LSB absolute accuracy,
- 13 ADC clock cycle conversion time,

- 8 multiplexed single-ended input channels,
- selectable right or left result justification, and
- 0 to Vcc ADC input voltage range.

Let us discuss each feature in turn. The first feature of discussion is "10-bit resolution." Resolution is defined as:

$$\text{resolution} = (V_{RH} - V_{RL})/2^b.$$

V_{RH} and V_{RL} are the ADC high and low reference voltages, whereas b is the number of bits available for conversion. For the ATmega16 with reference voltages of 5 VDC, 0 VDC, and 10 bits available for conversion, resolution is 4.88 mV. Absolute accuracy specified as ± 2 LSB is then ± 9.76 mV at this resolution.

It requires 13 analog-to-digital clock cycles to perform an ADC conversion. The ADC system may be run at a slower clock frequency than the main microcontroller clock source. The main microcontroller clock is divided down using the ADC Prescaler Select (ADPS[2:0]) bits in the ADC Control and Status Register A (ADCSRA).

The ADC is equipped with a single successive approximation converter. Only a single ADC channel may be converted at a given time. The input of the ADC is equipped with an eight-input analog multiplexer. The analog input for conversion is selected using the MUX[4:0] bits in the ADC Multiplexer Selection Register (ADMUX).

The 10-bit result from the conversion process is placed in the ADC Data Registers, ADCH and ADCL. These two registers provide 16 bits for the 10-bit result. The result may be left justified by setting the ADLAR (ADC Left Adjust Result) bit of the ADMUX register. Right justification is provided by clearing this bit.

The analog input voltage for conversion must be between 0 and Vcc V. If this is not the case, external circuitry must be used to ensure the analog input voltage is within these prescribed bounds as discussed earlier in the chapter.

3.4.1 Block Diagram

The block diagram for the ATmega16 ADC conversion system is provided in Figure 3.9. The left edge of the diagram provides the external microcontroller pins to gain access to the ADC. The eight analog input channels are provided at ADC[7:0], and the ADC reference voltage pins are provided at AREF and AVCC. The key features and registers of the ADC system previously discussed are included in the diagram.

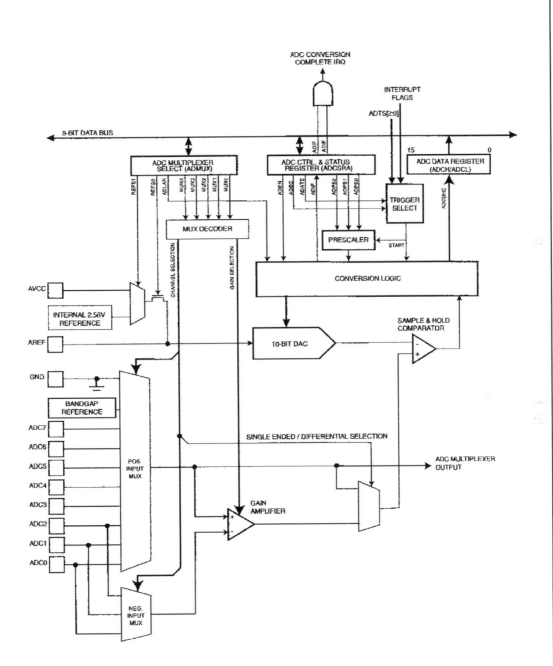

FIGURE 3.9: Atmel AVR ATmega16 ADC block diagram. Figure used with permission of Atmel.

3.4.2 Registers

The key registers for the ADC system are shown in Figure 3.10. It must be emphasized that the ADC system has many advanced capabilities that we do not discuss here. Our goal is to review the basic ADC conversion features of this powerful system. We have already discussed many of the register settings already. We will discuss each register in turn.

3.4.2.1 ADC Multiplexer Selection Register. As previously discussed, the ADMUX register contains the ADLAR bit to select left or right justification and the MUX[4:0] bits to determine which analog input will be provided to the ADC for conversion. To select a specific input for conversion is accomplished when a binary equivalent value is loaded into the MUX[4:0] bits. For example, to convert channel ADC7, "00111" is loaded into the ADMUX register. This may be accomplished using the following C instruction:

```
ADMUX = 0x07;
```

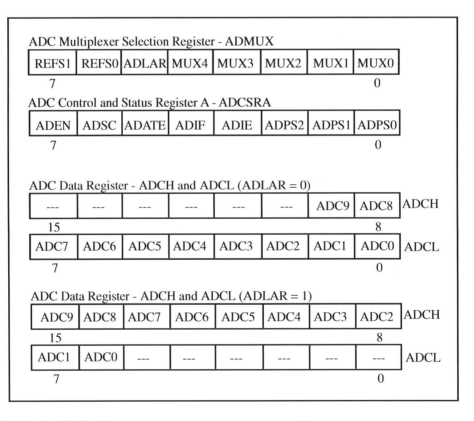

FIGURE 3.10: ADC registers.

The REFS[1:0] bits of the ADMUX register are also used to determine the reference voltage source for the ADC system. These bits may be set to the following values:

- REFS[0:0] = 00: AREF used for ADC voltage reference
- REFS[0:1] = 01: AVCC with external capacitor at the AREF pin
- REFS[1:0] = 10: reserved
- REFS[1:1] = 11: internal 2.56-VDC voltage reference with an external capacitor at the AREF pin

3.4.2.2 ADC Control and Status Register A. The ADCSRA register contains the ADC Enable (ADEN) bit. This bit is the "on/off" switch for the ADC system. The ADC is turned on by setting this bit to a logic 1. The ADC Start Conversion (ADSC) bit is also contained in the ADCSRA register. Setting this bit to logic 1 initiates an ADC. The ADCSRA register also contains the ADC Interrupt flag (ADIF) bit. This bit sets to logic 1 when the ADC is complete. The ADIF bit is reset by writing a logic 1 to this bit.

The ADPS[2:0] bits are used to set the ADC clock frequency. The ADC clock is derived from dividing down the main microcontroller clock. The ADPS[2:0] may be set to the following values:

- ADPS[2:0] = 000: division factor: 2
- ADPS[2:0] = 001: division factor: 2
- ADPS[2:0] = 010: division factor: 4
- ADPS[2:0] = 011: division factor: 8
- ADPS[2:0] = 100: division factor: 16
- ADPS[2:0] = 101: division factor: 32
- ADPS[2:0] = 110: division factor: 64
- ADPS[2:0] = 111: division factor: 128

3.4.2.3 ADC Data Registers (ADCH and ADCL). As previously discussed, the ADC Data Register contains the result of the ADC. The results may be left (ADLAR=1) or right (ADLAR=0) justified.

3.4.3 Programming the ADC

Provided below are two functions to operate the ATmega16 ADC system. The first function "InitADC( )" initializes the ADC by first performing a dummy conversion on channel 0. In this particular example the ADC prescalar is set to 8 (the main microcontroller clock was operating at 10 MHz).

The function then enters a while loop waiting for the ADIF bit to set, indicating the conversion is complete. After conversion, the ADIF bit is reset by writing a logic 1 to it.

The second function "ReadADC(unsigned char)" is used to read the analog voltage from the specified ADC channel. The desired channel for conversion is passed in as an unsigned character variable. The result is returned as a left-justified, 10-bit binary result. The ADC prescalar must be set to 8 to slow down the ADC clock at higher external clock frequencies (10 MHz) to obtain accurate results. After the ADC is complete, the results in the 8-bit ADCL and ADCH result registers are concatenated into a 16-bit unsigned integer variable and returned to the function call. In the final chapter of the book, we show how to convert an unsigned integer into a floating point voltage that may be displayed on a LCD.

```
//*************************************************************
//InitADC: initialize analog-to-digital converter
//*************************************************************

void InitADC( void)
{
ADMUX = 0;                          //Select channel 0
ADCSRA = 0xC3;                      //Enable ADC & start 1st
                                    //dummy
                                    //conversion
                                    //Set ADC module prescalar
                                    //to 8 critical for
                                    //accurate ADC results
while (!(ADCSRA & 0x10));           //Check if conversation is
                                    //ready
ADCSRA |= 0x10;                     //Clear conv rdy flag -
                                    //set the bit

}


//*************************************************************
//ReadADC: read analog voltage from analog-to-digital converter -
//the desired channel for conversion is passed in as an unsigned
```

```c
//character variable. The result is returned as a left justified,
//10 bit binary result. The ADC prescalar must be set to 8 to
//slow down the ADC clock at higher external clock frequencies
//(10 MHz) to obtain accurate results.
//*************************************************************


unsigned int ReadADC(unsigned char channel)

{
unsigned int binary_weighted_voltage, binary_weighted_voltage_low;
unsigned int binary_weighted_voltage_high;  //weighted binary
                                            //voltage

ADMUX = channel;                        //Select channel
ADCSRA |= 0x43;                         //Start conversion
                                        //Set ADC module prescalar
                                        //to 8 critical for
                                        //accurate ADC results
while (!(ADCSRA & 0x10));               //Check if converstion is
                                        //ready
ADCSRA |= 0x10;                         //Clear Conv rdy flag - set
                                        //the bit
binary_weighted_voltage_low = ADCL;     //Read 8 low bits first
                                        //(important)
                                        //Read 2 high bits,
                                        //multiply by 256
binary_weighted_voltage_high = ((unsigned int)(ADCH << 8));
binary_weighted_voltage =
        = binary_weighted_voltage_low | binary_weighted_
            voltage_high;
return binary_weighted_voltage;         //ADCH:ADCL
}

//*************************************************************
```

3.4.4 Digital-to-Analog Conversion

Once a signal is acquired to a digital system with the help of the ADC process and has been processed, frequently, the processed signal is converted back to another analog signal. A simple example of such a conversion occurs in digital audio processing. Human voice is converted to a digital signal, modified, processed, and converted back to an analog signal for people to hear. The process to convert digital signals to analog signals is completed by a DAC. The most commonly used technique to convert digital signals to analog signals is the summation method shown in Figure 3.11.

With the summation method of DAC, a digital signal, represented by a set of 1's and 0's, enters the DAC from the MSB to the LSB. For each bit, a comparator checks its logic state, high or low, to produce a clean digital bit, represented by a voltage level. Typically, in a microcontroller

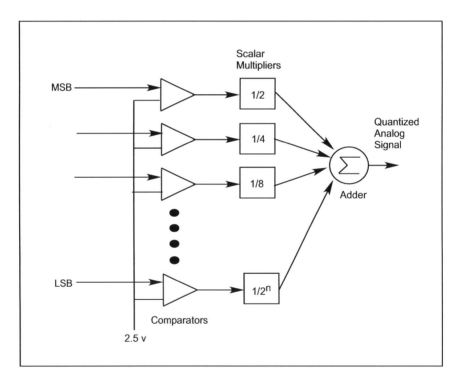

FIGURE 3.11: A summation method to convert a digital signal into a quantized analog signal. Comparators are used to clean up incoming signals, and the resulting values are multiplied by a scalar multiplier, and the results are added to generate the output signal. For the final analog signal, the quantized analog signal should be connected to a low pass filter followed by a transducer interface circuit.

context, the voltage level is 5 or 0 V to represent logic 1 or logic 0, respectively. The voltage is then multiplied by a scalar value based on its significant position of the digital signal as shown in Figure 3.11. Once all bits for the signal have been processed, the resulting voltage levels are summed to produce the final analog voltage value. Notice that the production of a desired analog signal may involve further signal conditioning such as a low pass filter to "smooth" the quantized analog signal and a transducer interface circuit to match the output of the DAC to the input of an output transducer.

3.5 SUMMARY

In this chapter, we presented the differences between analog and digital signals and used this knowledge to discuss three subprocessing steps involved in ADCs: sampling, quantization, and encoding. We also presented the quantization errors and the data rate associated with the ADC process. The dynamic range of an ADC, one of the measures to describe a conversion process, was also presented. Two different categories exist to represent technologies used to convert analog signals to their corresponding digital signals: direct approach and indirect approaches. For the direct approach, we presented the successive approximation, counter-based, and parallel conversion techniques. For the indirect approach, the integration-based conversion technique was discussed. We then reviewed the operation, registers, and actions required to program the ADC system aboard the ATmega16. We concluded the chapter with a brief presentation of DACs.

3.6 REFERENCES AND FURTHER READING

1. D Pack and S Barrett, *Microcontroller Theory and Applications: HC12 and S12*, 2nd ed, Prentice-Hall, Upper Saddle River, NJ, in press.
2. S Barrett and D Pack, *Microcontrollers Fundamentals for Engineers and Scientists*, Morgan & Claypool, San Rafael, CA, 2006. doi:10.2200/S00025ED1V01Y200605DCS001
3. A Oppenheim and Ronald Schafer, *Discrete-Time Signal Processing*, 2nd ed, Prentice-Hall, Upper Saddle River, NJ, 1999.
4. R Thomas and A Rosa, *The Analysis and Design of Linear Circuits*, 4th ed, Wiley & Sons, New York, 2003.
5. J Enderle, S Blanchard, and J Bronzino, *Introduction to Biomedical Engineering*, Academic Press, San Diego, 2000.
6. *Atmel 8-bit AVR Microcontroller with 16K Bytes In-System Programmable Flash, ATmega16, ATmega16L*, data sheet: 2466L-AVR-06/05, Atmel, San Jose, CA.

3.7 CHAPTER PROBLEMS

1. Question: Given a sinusoid with 500-Hz frequency, what should be the minimum sampling frequency for an ADC if we want to faithfully reconstruct the analog signal after the conversion?
2. Question: If 12 bits are used to quantize a sampled signal, what is the number of available quantized levels? What will be the resolution of such a system if the input range of the ADC is 10 V?
3. Question: Given the 12-V input range of an ADC and the desired resolution of 0.125 V, what should be the minimum number of bits used for the conversion?
4. Question: Perform a trade-off study on the four technologies used for the ADC. Use cost, conversion time, and accuracy as the list of criteria.
5. Question: Investigate the ADCs in your audio system. Find the sampling rate, the quantization bits, and the technique used for the conversion.
6. Question: A flex sensor provides 10 kΩ of resistance for 0deg[TS: I can't generate the degree symbol.] flexure and 40 kΩ of resistance for 90deg[TS: I can't generate the degree symbol.] of flexure. Design a circuit to convert the resistance change to a voltage change (hint: consider a voltage divider). Then design a transducer interface circuit to convert the output from the flex sensor circuit to voltages suitable for the ATmega16 ADC system.
7. Question: If an analog signal is converted by an ADC to a binary representation and then back to an analog voltage using a DAC, will the original analog input voltage be the same as the resulting analog output voltage? Explain.

CHAPTER 4

Interrupt Subsystem

Objectives: After reading this chapter, the reader should be able to

- understand the need of a microcontroller for interrupt capability,
- describe the general microcontroller interrupt response procedure,
- describe the ATmega16 interrupt features, and
- properly configure and program an interrupt event for the ATmega16.

4.1   INTERRUPT THEORY

A microcontroller normally executes instructions in an orderly fetch–decode–execute sequence as dictated by a user-written program as shown in Figure 4.1. However, the microcontroller must be equipped to handle unscheduled, higher-priority events that might occur inside or outside the microcontroller. To process such events, a microcontroller requires an interrupt system [1].

The interrupt system onboard a microcontroller allows it to respond to higher-priority events. These events may be planned, but we do not know when they will occur. When an interrupt event occurs, the microcontroller will normally complete the instruction it is currently executing and then transition program control to interrupt event specific tasks. These tasks, which resolve the interrupt event, are organized into a function called an interrupt service routine (ISR). Each interrupt will normally have its own interrupt specific ISR. Once the ISR is complete, the microcontroller will resume processing where it left off before the interrupt event occurred.

4.2   ATmega16 INTERRUPT SYSTEM

The ATmega16 is equipped to handle a powerful and flexible complement of 21 interrupt sources. Three of the interrupts originate from external interrupt sources, whereas the remaining 18 interrupts support the efficient operation of peripheral subsystems aboard the microcontroller. The ATmega16 interrupt sources are shown in Figure 4.2. The interrupts are listed in descending order of priority. As you can see, the RESET has the highest priority, followed by the external interrupt request pins INT0 (pin 16) and INT1 (pin 17). The remaining interrupt sources are internal to the ATmega16.

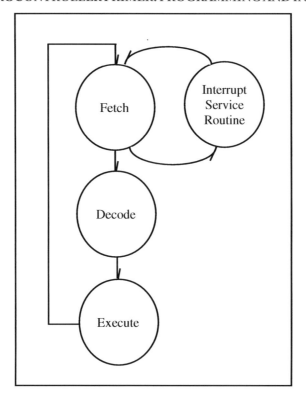

FIGURE 4.1: Microcontroller interrupt response.

When an interrupt occurs, the microcontroller completes the current instruction, stores the address of the next instruction on the stack, and starts executing instructions in the designated ISR corresponding to the particular interrupt source. It also turns off the interrupt system to prevent further interrupts while one is in progress. The execution of the ISR is performed by loading the beginning address of the ISR specific for that interrupt into the program counter. The ISR will then commence. Execution of the ISR continues until the return from interrupt instruction (reti) is encountered. Program control then reverts back to the main program.

4.3 PROGRAMMING AN INTERRUPT

To program an interrupt, the user is responsible for the following actions:

- Ensure the ISR for a specific interrupt is tied to the correct interrupt vector address, which points to the starting address of the ISR.

INTERRUPT SUBSYSTEM 67

Vector No.	Program Address[2]	Source	Interrupt Definition
1	$000[1]	RESET	External Pin, Power-on Reset, Brown-out Reset, Watchdog Reset, and JTAG AVR Reset
2	$002	INT0	External Interrupt Request 0
3	$004	INT1	External Interrupt Request 1
4	$006	TIMER2 COMP	Timer/Counter2 Compare Match
5	$008	TIMER2 OVF	Timer/Counter2 Overflow
6	$00A	TIMER1 CAPT	Timer/Counter1 Capture Event
7	$00C	TIMER1 COMPA	Timer/Counter1 Compare Match A
8	$00E	TIMER1 COMPB	Timer/Counter1 Compare Match B
9	$010	TIMER1 OVF	Timer/Counter1 Overflow
10	$012	TIMER0 OVF	Timer/Counter0 Overflow
11	$014	SPI, STC	Serial Transfer Complete
12	$016	USART, RXC	USART, Rx Complete
13	$018	USART, UDRE	USART Data Register Empty
14	$01A	USART, TXC	USART, Tx Complete
15	$01C	ADC	ADC Conversion Complete
16	$01E	EE_RDY	EEPROM Ready
17	$020	ANA_COMP	Analog Comparator
18	$022	TWI	Two-wire Serial Interface
19	$024	INT2	External Interrupt Request 2
20	$026	TIMER0 COMP	Timer/Counter0 Compare Match
21	$028	SPM_RDY	Store Program Memory Ready

FIGURE 4.2: Atmel AVR ATmega16 Interrupts [2]. Figure used with permission of Atmel.

- Ensure the interrupt system has been globally enabled. This is accomplished with the assembly language instruction SEI.
- Ensure the specific interrupt subsystem has been locally enabled.
- Ensure the registers associated with the specific interrupt have been configured correctly.

In the next two examples that follow, we illustrate how to accomplish these steps. We use the ImageCraft ICC AVR compiler, which contains excellent support for interrupts. Other compilers have similar features.

4.4 APPLICATION

In this section, we provide two representative samples of writing interrupts for the ATmega16. We provide both an externally generated interrupt event and also one generated from within the microcontroller. The ImageCraft ICC AVR compiler uses the following syntax to link an ISR to the correct interrupt vector address:

```
#pragma interrupt_handler timer_handler:4

void timer_handler(void)
{
:
:
}
```

As you can see, the #pragma with the reserved word **interrupt_handler** is used to communicate to the compiler that the routine name that follows is an ISR. The number that follows the ISR name corresponds to the interrupt vector number in Figure 4.2. The ISR is then written like any other function. It is important that the ISR name used in the #pragma instruction matches the name of the ISR in the function body. Because the compiler knows the function is an ISR, it will automatically place the RETI instruction at the end of the ISR when the corresponding assembly language is generated.

4.4.1 External Interrupts

The external interrupts INT0 (pin 16), INT1 (pin 17), and INT2 (pin 3) trigger an interrupt within the ATmega16 when an external event of user-specified characteristics occurs at the pin associated with the specific interrupt. Interrupts INT0 and INT1 may be triggered with a level or an edge signal, whereas interrupt INT2 is edge-triggered only. The specific settings for each interrupt is provided in Figure 4.3.

Provided below is the code snapshot to configure an interrupt for INT0. In this specific example, an interrupt will occur when a positive edge transition occurs on the ATmega16 INT0 external interrupt pin.

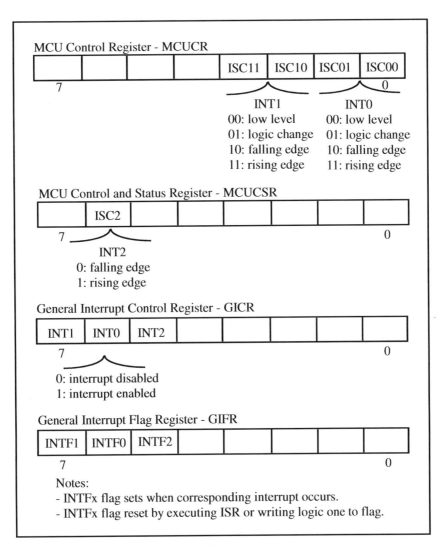

FIGURE 4.3: Interrupt INT0, INT1, and INT2 registers.

```c
//interrupt handler definition
#pragma interrupt_handler int0_ISR:2

//function prototypes
void int0_ISR(void);
void initialize_interrupt0(void);


//****************************************************************

//The following function call should be inserted in the main
//program to initialize the INT0 interrupt to respond to a
//positive edge trigger.
//This function should only be called once.

    :
initialize_interrupt0();
    :


//****************************************************************
//function definitions

//****************************************************************
//initialize_interrupt0:  initializes interrupt INT0.
//Note: stack is automatically initialized by the compiler
//****************************************************************

void initialize_interrupt0(void)        //initialize interrupt 0
{
DDRD = 0xFB;                            //set PD2 (int0) as input
PORTD &= ~0x04;                         //disable pullup resistor
                                        //PD2

GICR = 0x40;                            //enable int 0
MCUCR = 0x03;                           //set for positive edge
                                        //trigger

asm("SEI");                             //global interrupt enable
}
```

```
//**********************************************************
//int0_ISR: interrupt service routine for INT0
//**********************************************************

void int0_ISR(void)
{

//Insert interrupt specific actions here.

}
```

4.4.2 Internal Interrupt

In this example, we use Timer/Counter0 to provide prescribed delays within our program. Recall that Timer/Counter0 is an 8-bit timer. It rolls over every time it receives 256 timer clock "ticks." There is an interrupt associated with the Timer/Counter0 Overflow. If activated, the interrupt will occur every time the contents of the Timer/Counter0 transitions from 255 back to 0 count. We can use this overflow interrupt as a method of keeping track of real clock time (hours, minutes, and seconds) within a program. In this specific example, we use the overflow to provide precision program delays.

In this example, the ATmega16 is being externally clocked by a 10-MHz ceramic resonator. The resonator frequency is further divided by 256 using the CS[2:1:0] bits in the Timer/Counter Control Register 0 (TCCR0). When CS[2:1:0] are set for [1:0:0], the incoming clock source is divided by 256. This provides a clock tick to Timer/Counter0 every 25.6 μs. Therefore, the 8-bit Timer/Counter0 will rollover every 256 clock ticks, or every 6.55 ms.

To create a precision delay, we write a function called delay. The function requires an unsigned integer parameter value indicating how many 6.55-ms interrupts the function should delay. The function stays within a while loop until the desired number of interrupts has occurred. For example, to delay 1 s, the function would be called with the parameter value "153." That is, it requires 153 interrupts occurring at 6.55-ms intervals to generate a 1-s delay.

The code snapshots to configure the Time/Counter0 Overflow interrupt is provided below along with the associated ISR and the delay function.

```
//function prototypes******************************************
                                        //delay specified number
                                        //6.55ms
void delay(unsigned int number_of_6_55ms_interrupts);
```

```c
void init_timer0_ovf_interrupt(void);    //initialize timer0
                                         //overflow interrupt

//interrupt handler definition**********************************
                                         //interrupt handler
                                         //definition
#pragma interrupt_handler timer0_interrupt_isr:10

//global variables**********************************************
unsigned int   input_delay;              //counts number of
                                         //Timer/Counter0
                                         //Overflow interrupts



//main program**************************************************

void main(void)
{
init_timer0_ovf_interrupt();             //initialize Timer/
                                         //Counter0
                                         //Overflow
                                         //interrupt □ call once at
                                         //beginning of program
:
:
delay(153);                              //1 second delay

}
//***************************************************************
//int_timer0_ovf_interrupt(): The Timer/Counter0 Overflow
//interrupt is being employed as a time base for a master
//timer for this project.
//The ceramic resonator operating at 10 MHz is divided by 256.
//The 8-bit Timer0 register (TCNT0) overflows every 256 counts or
//every 6.55 ms.
//***************************************************************
```

INTERRUPT SUBSYSTEM

```c
void init_timer0_ovf_interrupt(void)
{
TCCR0 = 0x04;   //divide timer0 timebase by 256, overflow occurs
                //every 6.55ms
TIMSK = 0x01;   //enable timer0 overflow interrupt
asm("SEI");     //enable global interrupt
}




//****************************************************************
//****************************************************************
//timer0_interrupt_isr:
//Note: Timer overflow 0 is cleared by hardware when executing
//       the corresponding interrupt handling vector.
//****************************************************************

void timer0_interrupt_isr(void)
{
input_delay++;                          //increment overflow
                                        //counter

}



//****************************************************************
//delay(unsigned int num_of_6_55ms_interrupts): this generic delay
//function provides the specified delay as the number of 6.55 ms
//"clock ticks" from the Timer/Counter0 Overflow interrupt.
//Note: this function is only valid when using a 10 MHz crystal or
//       ceramic resonator
//****************************************************************

void delay(unsigned int number_of_6_55ms_interrupts)
{
TCNT0 = 0x00;                           //reset timer0
input_delay = 0;                        //reset timer0 overflow
                                        //counter
```

```
while(input_delay <= number_of_6_55ms_interrupts)
  {
  ;                                         //wait for specified
                                            //number of

  }                                         //interrupts
}

//*************************************************************
```

4.5 SUMMARY

In this chapter, we provided an introduction to the interrupt features available aboard the ATmega16. We also discussed how to program an interrupt for proper operation and provided two representative samples: an external and an internal interrupt.

4.6 REFERENCES AND FURTHER READING

1. S Barrett and D Pack, *Microcontrollers Fundamentals for Engineers and Scientists*, Morgan & Claypool, San Rafael, CA, 2006. doi:10.2200/S00025ED1V01Y200605DCS001
2. *Atmel 8-bit AVR Microcontroller with 16K Bytes In-System Programmable Flash, ATmega16, ATmega16L*, data sheet: 2466L-AVR-06/05, Atmel, San Jose, CA.

4.7 CHAPTER PROBLEMS

1. Question: What is the purpose of an interrupt?
2. Question: Describe the flow of events when an interrupt occurs.
3. Question: Describe the interrupt features available with the ATmega16.
4. Question: What is interrupt priority? How is it determined?
5. Question: What steps are required by the system designer to properly configure an interrupt?
6. Question: How is the interrupt system turned "on" and "off"?
7. Question: A 10-MHz ceramic resonator is not available. Redo the example of the Timer/Counter0 Overflow interrupt provided with a time base of 1 and 8 MHz.
8. Question: What is the maximum delay that may be generated with the delay function provided in the text without modification? How could the function be modified for longer delays?
9. Question: Develop a 24-hour timer (hh:mm:ss) using the Timer/Counter0 Overflow interrupt. What is the accuracy of your timer? How can it be improved?

CHAPTER 5
Timing Subsystem

Objectives: After reading this chapter, the reader should be able to

- describe key timing system-related terminology,
- compute the frequency and the period of a periodic signal using a microcontroller,
- explain functional components of a microcontroller timer system,
- describe the procedure to capture incoming signal events,
- describe the procedure to generate time critical output signals,
- describe the timing-related features of the Atmel ATmega16,
- describe the four operating modes of the Atmel ATmega16 timing system,
- describe the register configurations for the ATmega16's Timer 0, Timer 1, and Timer 2, and
- program the ATmega16 timer system.

5.1 OVERVIEW

One of the most important reasons for using microcontrollers in embedded systems is the capabilities of microcontrollers to perform time-related tasks. In a simple application, one can program a microcontroller system to turn on or turn off an external device at a programmed time. In a more involved application, we can use a microcontroller to generate complex digital waveforms with varying pulse widths to control the speed of a DC motor [1]. In this chapter, we review the capabilities of the Atmel ATmega16 [2] microcontroller to perform time-related functions. We begin with a review of timing-related terminology. We then provide an overview of the general operation of a timing system followed by the timing system features aboard the ATmega16. Next, we present a detailed discussion of each of its timing channels, Timer 0, Timer 1, and Timer 2, and their different modes of operation.

5.2 TIMING-RELATED TERMINOLOGY

5.2.1 Frequency

Consider signal $x(t)$ that repeats itself. We call this signal periodic with period T if it satisfies

$$x(t) = x(t + T).$$

To measure the frequency of a periodic signal, we count the number of times a particular event repeats within a 1-s period. The unit of frequency is Hertz, or cycles per second. For example, a sinusoidal signal with a 60-Hz frequency means that a full cycle of a sinusoid signal repeats itself 60 times each second, or every 16.67 ms.

5.2.2 Period

The flip side of a frequency is a period. If an event occurs with a rate of 1 Hz, the period of that event is 1 s. To find a period, given a frequency, or vice versa, we simply need to remember their inverse relationship $f = \frac{1}{T}$ where f and T represent a frequency and the corresponding period, respectively. Both periods and frequencies of signals are often used to specify timing constraints of embedded systems [3,4]. For example, when your car is on a wintery road and slipping, the engineers who designed your car configured the antislippage unit to react within some millisecond period, say 20 ms. The constraint then forces the design team that monitors the slippage to program their monitoring system to check a slippage at a rate of 50 Hz.

5.2.3 Duty Cycle

In many applications, periodic pulses are used as control signals. A good example is the use of a periodic pulse to control a servo motor. To control the direction and sometimes the speed of a motor, a periodic pulse signal with a changing duty cycle over time is used. The periodic pulse signal shown in Figure 5.1(a) is on for 50% of the signal period and off for the rest of the period. The pulse shown in Figure 5.1(b) is on for only 25% of the same period as the signal in Figure 5.1(a) and off for 75% of the period. The duty cycle is defined as the percentage of one period a signal is on. Therefore, we call the signal in Figure 5.1(a) the periodic pulse signal with a 50% duty cycle and the corresponding signal in Figure 5.1(b) the periodic pulse signal with a 25% duty cycle.

5.3 TIMING SYSTEM OVERVIEW

The heart of the timing system is the crystal time base. The crystal's frequency of a microcontroller is used to generate a baseline clock signal. For a timer system, the system clock is used to update the contents of a special register called a free-running counter. The job of a free-running counter is to count (increment) each time it sees a rising edge (or a falling edge) of a clock signal. Thus, if a clock is running at the rate of 2 MHz, the free-running counter will count each 0.5 μs. All other

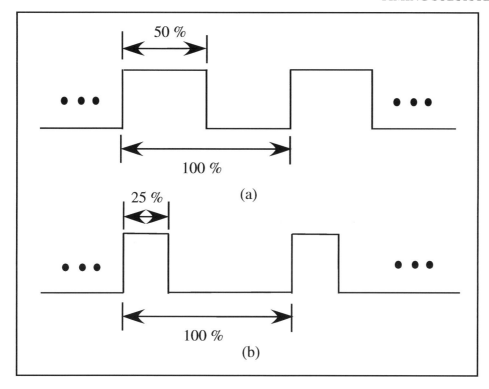

FIGURE 5.1: Two signals with the same period but different duty cycles: (a) periodic signal with a 50% duty cycle and (b) periodic signal with a 25% duty cycle.

timer-related units reference the contents of the free-running counter to perform I/O time-related activities: measurement of periods, capture of timing events, and generation of time-related signals.

For input time-related activities, all microcontrollers typically have timer hardware components that detect signal logic changes on one or more input pins. Such components rely on a free-running counter to capture external event times. We can use such ability to measure the period of an incoming signal, the width of a pulse, and the time of a signal logic change.

For output timer functions, a microcontroller uses a comparator, a free-running counter, logic switches, and special-purpose registers to generate time-related signals on one or more output pins. A comparator checks the value of the free-running counter for a match with the contents of another special-purpose register where a programmer stores a specified time in terms of the free-running counter value. The checking process is executed at each clock cycle, and when a match occurs, the corresponding hardware system induces a programmed logic change on a programmed

78 ATMEL AVR MICROCONTROLLER PRIMER: PROGRAMMING AND INTERFACING

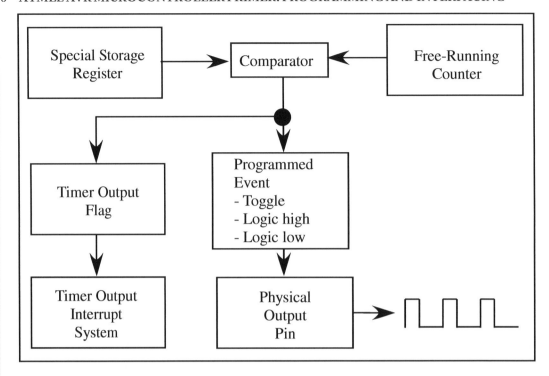

FIGURE 5.2: A diagram of a timer output system.

output port pin [5]. Using such capability, one can generate a simple logic change at a designated time incident, a pulse with a desired time width, or a PWM signal to control servo or DC motors.

You can also use the timer input system to measure the pulse width of an aperiodic signal. For example, suppose that the times for the rising edge and the falling edge of an incoming signal are 1.5 and 1.6 s, respectively. We can use these values to easily compute the pulse width of 0.1 s.

The second overall goal of the timer system is to generate signals to control external devices. Again, an event simply means a change of logic states on an output pin of a microcontroller at a specified time. Now consider Figure 5.2. Suppose an external device connected to the microcontroller requires a pulse signal to turn itself on. Suppose the particular pulse the external device needs is 2-ms wide. In such situations, we can use the free-running counter value to synchronize the time of desired logic state changes. Naturally, extending the same capability, we can also generate a periodic pulse with a fixed duty cycle or a varying duty cycle.

From the examples we discussed above, you may have wondered how a microcontroller can be used to compute absolute times from the relative free-running counter values, say 1.5 and

1.6 s. The simple answer is that we cannot do so directly. A programmer must use the relative system clock values and derive the absolute time values. Suppose your microcontroller is clocked by a 2-MHz signal, and the system clock uses a 16-bit free-running counter. For such a system, each clock period represents 0.5 μS, and it takes approximately 32.78 ms to count from 0 to 2^{16} (65,536). The timer input system then uses the clock values to compute frequencies, periods, and pulse widths. For example, suppose you want to measure a pulse width of an incoming aperiodic signal. If the rising edge and the falling edge occurred at count values $0010 and $0114, can you find the pulse width when the free-running counter is counting at 2 MHz? Recall that the $ symbol represents that the following value is in a hexadecimal form. Let us first convert the two values into their corresponding decimal values, 276 and 16. The pulse width of the signal in the number of counter value is 260. Because we already know how long it takes for the system to count 1, we can readily compute the pulse width as $260 \times 0.5\ \mu s = 130\ \mu s$.

Our calculations do not take into account time increments lasting longer than the rollover time of the counter. When a counter rolls over from its maximum value back to 0, a flag is set to notify the processor of this event. The rollover events may be counted to correctly determine the overall elapsed time of an event.

5.4 APPLICATIONS

In this section, we consider some important uses of the timer system of a microcontroller to (1) measure an input signal timing event, termed input capture, (2) count the number of external signal occurrences, (3) generate timed signals—termed *output compare*, and, finally, (4) generate PWM signals. We first start with a case of measuring the time duration of an incoming signal.

5.4.1 Input Capture—Measuring External Timing Event

In many applications, we are interested in measuring the elapsed time or the frequency of an external event using a microcontroller. Using the hardware and functional units discussed in the previous sections, we now present a procedure to accomplish the task of computing the frequency of an incoming periodic signal. Figure 5.3 shows an incoming periodic signal to our microcontroller.

The first necessary step for the current task is to turn on the timer system. To reduce power consumption, a microcontroller usually does not turn on all of its functional systems after reset until they are needed. In addition to a separate timer module, many microcontroller manufacturers allow a programmer to choose the rate of a separate timer clock that governs the overall functions of a timer module.

Once the timer is turned on and the clock rate is selected, a programmer must configure the physical port to which the incoming signal arrives. This step is done using a special input timer port configuration register. The next step is to program the input event to capture. In our current

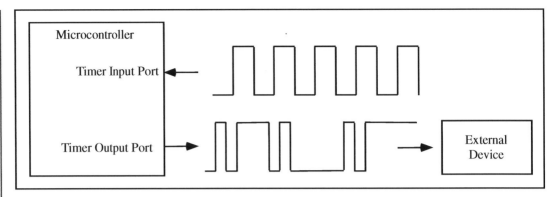

FIGURE 5.3: Use of the timer I/O systems of a microcontroller. The signal on top is fed into a timer input port. The captured signal is subsequently used to compute the input signal frequency. The signal on the bottom is generated using the timer output system. The signal is used to control an external device.

example, we should capture two consecutive rising edges or falling edges of the incoming signal. Again, the programming portion is done by storing an appropriate setup value to a special register.

Now that the input timer system is configured appropriately, you now have two options to accomplish the task. The first one is the use of a polling technique; the microcontroller continuously polls a flag, which holds a logic high signal when a programmed event occurs on the physical pin. Once the microcontroller detects the flag, it needs to clear the flag and record the time when the flag was set using another special register that captures the time of the associated free-running counter value. The program needs to continue to wait for the next flag, which indicates the end of one period of the incoming signal. A programmer then needs to record the newly acquired captured time represented in the form of a free-running counter value again. The period of the signal can now be computed by computing the time difference between the two captured event times, and based on the clock speed of the microcontroller, the programmer can compute the actual time changes and consequently the frequency of the signal.

In many cases, a microcontroller cannot afford the time to poll for one event. Such situation introduces the second method: interrupt systems. Most microcontroller manufacturers have developed built-in interrupt systems with their timer input modules. Instead of continuously polling for a flag, a microcontroller performs other tasks and relies on its interrupt system to detect the programmed event. The task of computing the period and the frequency is the same as the first method, except that the microcontroller will not be tied down constantly checking the flag, increasing the efficient use of the microcontroller resources. To use interrupt systems, of course, we must pay the price by appropriately configuring the interrupt systems to be triggered when a desired event is detected. Typically, additional registers must be configured, and a special program called an ISR must be written.

Suppose that for an input capture scenario, the two captured times for the two rising edges are $1000 and $5000, respectively. Note that these values are not absolute times but the representations of times reflected as the values of the free-running counter. The period of the signal is $4000, or 16384 in a decimal form. If we assume that the timer clock runs at 10 MHz, the period of the signal is 1.6384 ms, and the corresponding frequency of the signal is approximately 610.35 Hz.

5.4.2 Counting Events

The same capability of measuring the period of a signal can also be used to simply count external events. Suppose we want to count the number of logic state changes of an incoming signal for a given period. Again, we can use the polling technique or the interrupt technique to accomplish the task. For both techniques, the initial steps of turning on a timer and configuring a physical input port pin are the same. In this application, however, the programmed event should be any logic state changes instead of looking for a rising or a falling edge as we have done in the previous section. If the polling technique is used, at each event detection, the corresponding flag must be cleared and a counter must be updated. If the interrupt technique is used, one must write an ISR within which the flag is cleared and a counter is updated.

5.4.3 Output Compare—Generating Timing Signals to Interface External Devices

In the previous two sections, we considered two applications of capturing external incoming signals. In this subsection and the next one, we consider how a microcontroller can generate time critical signals for external devices. Suppose in this application, we want to send a signal shown in Figure 5.3 to turn on an external device. The timing signal is arbitrary, but the application will show that a timer output system can generate any desired time-related signals permitted under the timer clock speed limit of the microcontroller.

Similar to the use of the timer input system, one must first turn on the timer system and configure a physical pin as a timer output pin using special registers. In addition, one also needs to program the desired external event using another special register associated with the timer output system. To generate the signal shown in Figure 5.3, one must compute the time required between the rising and the falling edges. Suppose that the external device requires a pulse that is 2 ms wide to be activated. To generate the desired pulse, one must first program the logic state for the particular pin to be low and set the time value using a special register with respect to the contents of the free-running counter. As was mentioned in Section 5.2, at each clock cycle, the special register contents are compared with the contents of the free-running counter, and when a match occurs, the programmed logic state appears on the designated hardware pin. Once the rising edge is generated, the program then must reconfigure the event to be a falling edge (logic state low) and change the contents of the special register to be compared with the free-running counter. For the particular example in Figure 5.3, let us assume that the main clock runs at 2 MHz, the free-running counter is a 16-bit counter, and the name of the special register (16-bit register) where we can put appropriate values is output timer register. To generate the desired pulse, we can put $0000 first to the output timer register, and after the rising edge has been generated, we need to change the program event to a falling edge and put $0FA0 or 4000 in decimal to the output timer register. As was the case with the input timer system module, we can use output timer system interrupts to generate the desired signals as well.

5.4.4 Industrial Implementation Case Study (PWM)

In this section, we discuss a well-known method to control the speed of a DC motor using a PWM signal. The underlying concept is as follows. If we turn on a DC motor and provide the required voltage, the motor will run at its maximum speed. Suppose we turn the motor on and off rapidly by applying a periodic signal. The motor at some point cannot react fast enough to the changes of the voltage values and will run at the speed proportional to the average time the motor was turned on. By changing the duty cycle, we can control the speed of a DC motor as we desire. Suppose again we want to generate a speed profile shown in Figure 5.4. As shown in the figure, we want to accelerate the speed, maintain the speed, and decelerate the speed for a fixed amount of time.

The first task necessary is again to turn on the timer system, configure a physical port, and program the event to be a rising edge. As a part of the initialization process, we need to put $0000 to the output timer register we discussed in the previous subsection. Once the rising edge is generated, the program then needs to modify the event to a falling edge and change the contents of the output timer register to a value proportional to a desired duty cycle. For example, if we want to start off with 25% duty cycle, we need to input $4000 to the register, provided that we are using a 16-bit free-running counter. Once the falling edge is generated, we now need to go back and change the event to be a rising edge and the contents of the output timer counter value back to $0000. If we want to continue to generate a 25% duty cycle signal, then we must repeat the process indefinitely. Note that we are using the time for a free-running counter to count from $0000 to $FFFF as one period.

Now suppose we want to increase the duty cycle to 50% over 1 s and that the clock is running at 2 MHz. This means that the free-running counter counts from $0000 to $FFFF every 32.768 ms, and the free-running counter will count from $0000 to $FFFF approximately 30.51 times over the period of 1 s. That is, we need to increase the pulse width from $4000 to $8000 in approximately 30 turns, or approximately 546 clock counts every turn.

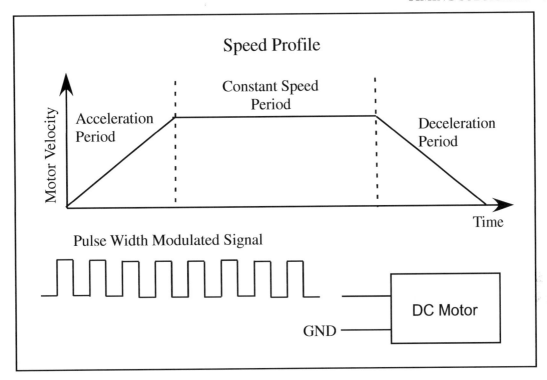

FIGURE 5.4: The figure shows the speed profile of a DC motor over time when a pulse-width-modulated signal is applied to the motor.

5.5 OVERVIEW OF THE ATMEL TIMERS

The Atmel ATmega16 is equipped with a flexible and powerful three-channel timing system. The timer channels are designated Timer 0, Timer 1, and Timer 2. In this section, we review the operation of the timing system in detail. We begin with an overview of the timing system features, followed by a detailed discussion of timer channel 0. Space does not permit a complete discussion of the other two timing channels; we review their complement of registers and highlight their features not contained in our discussion of timer channel 0. The information provided on timer channel 0 is readily adapted to the other two channels.

The features of the timing system are summarized in Figure 5.5. Timer 0 and Timer 2 are 8-bit timers, whereas Timer 1 is a 16-bit timer. Each timing channel is equipped with a prescaler. The prescaler is used to subdivide the main microcontroller clock source (designated $f_{clk\_I/O}$ in upcoming diagrams) down to the clock source for the timing system (clk_{Tn}).

Timer 0	Timer 1	Timer 2
- 8-bit timer/counter	- 16-bit timer/counter	- 8-bit timer/counter
- 10-bit clock prescaler	- 10-bit clock prescaler	- 10-bit clock prescaler
- Functions:	- Functions:	- Functions:
– Pulse width modulation	– Pulse width modulation	– Pulse width modulation
– Frequency generation	– Frequency generation	– Frequency generation
– Event counter	– Event counter	– Event counter
– Output compare	– Output compare – 2 ch	– Output compare
- Modes of operation:	– Input capture	- Modes of operation:
– Normal	- Modes of operation:	– Normal
– Clear timer on compare match (CTC)	– Normal	– Clear timer on compare match (CTC)
– Fast PWM	– Clear timer on compare match (CTC)	– Fast PWM
– Phase correct PWM	– Fast PWM	– Phase correct PWM
	– Phase correct PWM	

FIGURE 5.5: Atmel timer system overview.

Each timing channel has the capability to generate PWM signals, generate a periodic signal with a specific frequency, count events, and generate a precision signal using the output compare channels. Additionally, Timer 1 is equipped with the Input Capture feature.

All of the timing channels may be configured to operate in one of four operational modes designated : Normal (Mode 0), Clear Timer on Compare Match (CTC) (Mode 1), Fast PWM (mode 2), and Phase Correct PWM (mode 3). We provide more information on these modes shortly.

5.6 TIMER 0 SYSTEM

In this section, we discuss the features, overall architecture, modes of operation, registers, and programming of Timer 0. This information may be readily adapted to Timer 1 and Timer 2.

A Timer 0 block diagram is shown in Figure 5.6. The clock source for Timer 0 is provided via an external clock source at the T0 pin (PB0) of the microcontroller. Timer 0 may also be clocked internally via the microcontroller's main clock ($f_{clk\_I/O}$). This clock frequency may be too rapid for many applications. Therefore, the timing system is equipped with a prescaler to subdivide the main clock frequency down to timer system frequency (clk$_{Tn}$). The clock source for Timer 0 is selected using the CS0[2:0] bits contained in the Timer/Control Register (TCCR0). The TCCR0 register

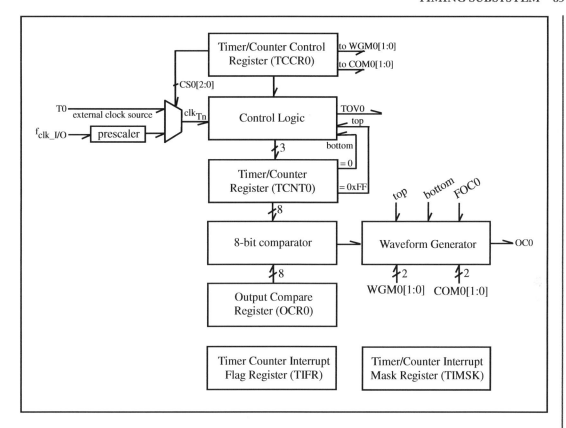

FIGURE 5.6: Timer 0 block diagram.

also contains the WGM0[1:0] and the COM0[1:0] bits, which are used to select the mode of operation for Timer 0 as well as tailor waveform generation for a specific application.

The timer clock source (clk_{T_n}) is fed to the 8-bit Timer/Counter Register (TCNT0). This register is incremented (or decremented) on each clk_{T_n} clock pulse. Timer 0 is also equipped with an 8-bit comparator that constantly compares the counts of TCNT0 to the Output Compare Register (OCR0). The compare signal from the 8-bit comparator is fed to the waveform generator. The waveform generator has a number of inputs (top, bottom, WGM0[1:0], and COM0[1:0]) to perform different operations with the timer system.

The BOTTOM signal for the waveform generation and the control logic, shown in Figure 5.6, is asserted when the timer counter TCNT0 reaches all 0's (0x00). The MAX signal for the control logic unit is asserted when the counter reaches all 1's (0xFF). The TOP signal for the waveform generation is asserted by either reaching the maximum count values of 0xFF on the

TCNT0 register or reaching the value set in the OCR0. The setting for the TOP signal will be determined by the timer's mode of operation.

Timer 0 also uses certain bits within the Timer/Counter Interrupt Mask Register (TIMSK) and the Timer/Counter Interrupt Flag Register (TIFR) to signal interrupt-related events. Timer 0 shares these registers with the other two timer channels.

5.6.1 Modes of Operation

Each of the timer channels may be set for a specific mode of operation: normal, CTC, fast PWM, and Phase Correct PWM. The system designer chooses the correct mode for the application at

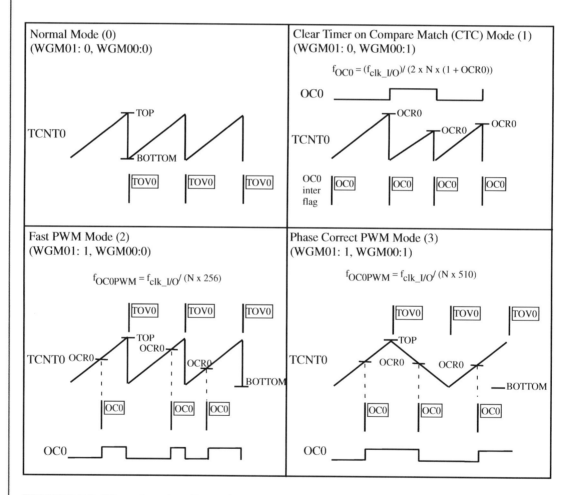

FIGURE 5.7: Timer 0 modes of operation.

TIMING SUBSYSTEM 87

hand. The timer modes of operation are summarized in Figure 5.7. A specific mode of operation is selected using the Waveform Generation Mode, WGM00 and WGM01, bits located in TCCR0.

5.6.1.1 Normal Mode. In the normal mode, the timer will continually count from 0x00 (BOTTOM) to 0xFF (TOP). When the TCNT0 returns to 0 on each cycle of the counter, the Timer/Counter Overflow Flag (TOV0) will be set.

The normal mode is useful for generating a periodic "clock tick" that may be used to calculate elapsed real time or provide delays within a system. We provide an example of this application in Section 5.9.

5.6.1.2 Clear Timer on Compare Match. In the CTC modem, the TCNT0 timer is reset to 0 every time the TCNT0 counter reaches the value set in OCR0. The Output Compare Flag 0 (OCF0) is set when this event occurs. The OCF0 flag is enabled by asserting the Timer/Counter 0 Output Compare Math Interrupt Enable (OCIE0) flag in the TIMSK and when the I-bit in the Status Register is set to 1.

The CTC mode is used to generate a precision digital waveform such as a periodic signal or a single pulse. The user must describe the parameters and key features of the waveform in terms of Timer 0 "clock ticks." When a specific key feature is reached within the waveform, the next key feature may be set into the OCR0 register.

5.6.1.3 Phase Correct PWM Mode. In the Phase Correct PWM mode, the TCNT0 register counts from 0x00 to 0xFF and back down to 0x00 continually. Every time the TCNT0 value matches the value set in the OCR0 register, the OCF0 flag is set and a change in the PWM signal occurs.

5.6.1.4 Fast PWM. The fast PWM mode is used to generate a precision PWM signal of a desired frequency and duty cycle. It is called the fast PWM because its maximum frequency is twice that of the Phase Correct PWM mode. When the TCNT0 register value reaches the value set in the OCR0 register, it will cause a change in the PWM output as prescribed by the system designer. It continues to count up to the TOP value, at which time the Timer/Counter 0 Overflow Flag (TOV0) is set.

5.6.2 Timer 0 Registers
A summary of the Timer 0 registers are shown in Figure 5.8.

88 ATMEL AVR MICROCONTROLLER PRIMER: PROGRAMMING AND INTERFACING

Timer/Counter Control Register (TCCR0)

FOC0	WGM00	COM01	COM00	WGM01	CS02	CS01	CS00
7							0

Timer/Counter Register (TCNT0)

7							0

Output Compare Register (OCR0)

7							0

Timer/Counter Interrupt Mask Register (TIMSK)

OCIE2	TOIE2	TICIE1	OCIE1A	OCIE1B	TOIE1	OCIE0	TOIE0
7							0

Timer/Counter Interrupt Flag REgister (TIFR)

OCF2	TOV2	ICF1	OCF1A	OCF1B	TOV1	OCF0	TOV0
7							0

FIGURE 5.8: Timer 0 registers.

5.6.2.1 Timer/Counter Control Register 0. The TCCR0 register bits are used to

- select the operational mode of Timer 0 using the Waveform Mode Generation (WGM0[1:0]) bits,
- determine the operation of the timer within a specific mode with the Compare Match Output Mode (COM0[1:0]) bits, and
- select the source of the Timer 0 clock using CS0[2:0] bits.

The bit settings for the TCCR0 register are summarized in Figure 5.9.

5.6.2.2 Timer/Counter Register. The TCNT0 is the 8-bit counter for Timer 0.

5.6.2.3 Output Compare Register. The OCR0 register holds a user-defined 8-bit value that is continuously compared with the TCNT0 register.

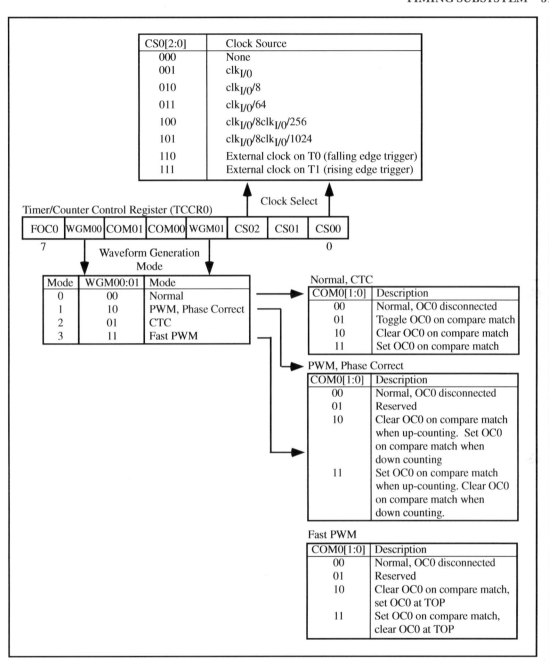

FIGURE 5.9: TCCR0 bit settings.

5.6.2.4 Timer/Counter Interrupt Mask Register. The TIMSK register is used by all three timer channels. Timer 0 uses the Timer/Counter 0 Output Compare Match Interrupt Enable (OCIE0) bit and the Timer/Counter 0 Overflow Interrupt Enable (TOIE0) bit. When the OCIE0 bit and the I-bit in the Status Register are both set to 1, the Timer/Counter 0 Compare Match interrupt is enabled. When the TOIE0 bit and the I-bit in the Status Register are both set to 1, the Timer/Counter 0 Overflow interrupt is enabled.

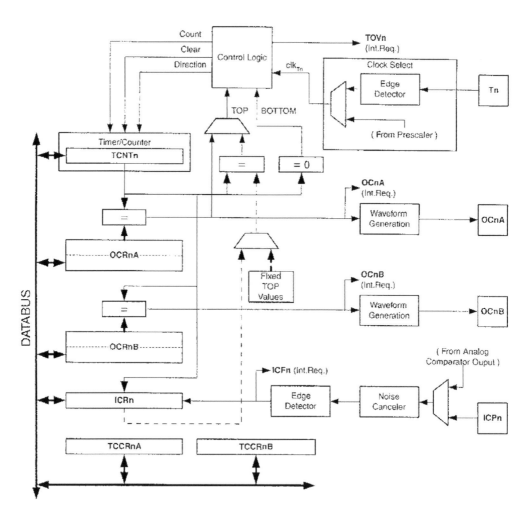

FIGURE 5.10: Timer 1 block diagram. Figure used with permission of Atmel.

5.6.2.5 Timer/Counter Interrupt Flag Register. The TIMSK register is used by all three timer channels. Timer 0 uses the OCF0, which sets for an output compare match. Timer 0 also uses the TOV0, which sets when Timer/Counter 0 Overflows.

5.7 TIMER 1

Timer 1 is a 16-bit timer/counter. It shares many of the same features of the Timer 0 channel. Because of limited space, the shared information will not be repeated. Instead, we concentrate on the enhancements of Timer 1, which include an additional output compare channel and also the capability for input capture. The block diagram for Timer 1 is shown in Figure 5.10.

As discussed earlier in the chapter, the input capture feature is used to capture the characteristics of an input signal, including period, frequency, duty cycle, or pulse length. This is accomplished by monitoring for a user-specified edge on the ICP1 microcontroller pin. When the desired edge occurs, the value of the Timer/Counter 1 (TCNT1) register is captured and stored in the Input Capture Register 1 (ICR1).

5.7.1 Timer 1 Registers

The complement of registers supporting Timer 1 are shown in Figure 5.11. Each register will be discussed in turn.

5.7.1.1 TCCR1A and TCCR1B Registers. The TCCR1 register bits are used to

- select the operational mode of Timer 1 using the Waveform Mode Generation (WGM1[3:0]) bits,
- determine the operation of the timer within a specific mode with the Compare Match Output Mode (Channel A, COM1A[1:0]; Channel B, COM1B[1:0]) bits, and
- select the source of the Timer 1 clock using CS1[2:0] bits.

The bit settings for the TCCR1A and TCCR1B registers are summarized in Figure 5.12.

5.7.1.2 Timer/Counter Register 1 (TCNT1H/TCNT1). The TCNT1 is the 16-bit counter for Timer 1.

Timer/Counter 1 Control Register A (TCCR1A)

COM1A1	COM1A0	COM1B1	COM1B0	FOC1A	FOC1B	WGM11	WGM10
7							0

Timer/Counter 1 Control Register B (TCCR1B)

ICNC1	ICES1	—	WGM13	WGM12	CS12	CS11	CS10
7							0

Timer Counter1 (TCNT1H/TCNT1L)

15							8
7							0

Output Compare Register 1 A (OCR1AH/OCR1AL)

15							8
7							0

Output Compare Register 1 B (OCR1BH/OCR1BL)

15							8
7							0

Input Capture Register 1 (ICR1H/ICR1L)

15							8
7							0

Timer/Counter Interrupt Mask Register (TIMSK)

OCIE2	TOIE2	TICIE1	OCIE1A	OCIE1B	TOIE1	OCIE0	TOIE0
7							0

Timer/Counter Interrupt Flag REgister (TIFR)

OCF2	TOV2	ICF1	OCF1A	OCF1B	TOV1	OCF0	TOV0
7							0

FIGURE 5.11: Timer 1 registers.

TIMING SUBSYSTEM 93

CS0[2:0]	Clock Source
000	None
001	$clk_{I/O}$
010	$clk_{I/O}/8$
011	$clk_{I/O}/64$
100	$clk_{I/O}/8clk_{I/O}/256$
101	$clk_{I/O}/8clk_{I/O}/1024$
110	External clock on T0 (falling edge trigger)
111	External clock on T1 (rising edge trigger)

Clock Select

Timer/Counter 1 Control Register B (TCCR1B)

ICNC1	ICES1	—	WGM13	WGM12	CS12	CS11	CS10
7							0

Timer/Counter 1 Control Register A (TCCR1A)

COM1A1	COM1A0	COM1B1	COM1B0	FOC1A	FOC1B	WGM11	WGM10
7							0

Waveform Generation Mode

Mode	WGM[13:12:11:10]	Mode
0	0000	Normal
1	0001	PWM, Phase Correct, 8-bit
2	0010	PWM, Phase Correct, 9-bit
3	0011	PWM, Phase Correct, 10-bit
4	0100	CTC
5	0101	Fast PWM, 8-bit
6	0110	Fast PWM, 9-bit
7	0111	Fast PWM, 10-bit
8	1000	PWM, Phase & Freq Correct
9	1001	PWM, Phase & Freq Correct
10	1010	PWM, Phase Correct
11	1011	PWM, Phase Correct
12	1100	CTC
13	1101	Reserved
14	1110	Fast PWM
15	1111	Fast PWM

Normal, CTC

COMx[1:0]	Description
00	Normal, OC1A/1B disconnected
01	Toggle OC1A/1B on compare match
10	Clear OC1A/1B on compare match
11	Set OC1A/1B on compare match

PWM, Phase Correct, Phase & Freq Correct

COMx[1:0]	Description
00	Normal, OC0 disconnected
01	WGM1[3:0] = 9 or 14: toggle OCnA on compare match, OCnB disconnected WGM1[3:0]= other settings, OC1A/1B disconnected
10	Clear OC0 on compare match when up-counting. Set OC0 on compare match when down counting
11	Set OC0 on compare match when up-counting. Clear OC0 on compare match when down counting.

Fast PWM

COMx[1:0]	Description
00	Normal, OC1A/1B disconnected
01	WGM1[3:0] = 15, toggle OC1A on compare match OC1B disconnected WGM1[3:0] = other settings, OC1A/1B disconnected
10	Clear OC1A/1B on compare match, set OC1A/1B at TOP
11	Set OC1A/1B on compare match, clear OC1A/1B at TOP

FIGURE 5.12: TCCR1A and TCCR1B registers.

5.7.1.3 Output Compare Register 1 Channel A (OCR1AH/ OCR1AL). The OCR1A register holds a user-defined 16-bit value that is continuously compared with the TCNT1 register when Channel A is used.

5.7.1.4 Output Compare Register 1 Channel B (OCR1BH/ OCR1BL). The OCR1B register holds a user-defined 16-bit value that is continuously compared with the TCNT1 register when Channel B is used.

5.7.1.5 Input Capture Register 1 (ICR1H/ICR1L). ICR1 is a 16-bit register used to capture the value of the TCNT1 register when a desired edge on ICP1 pin has occurred.

5.7.1.6 Timer/Counter Interrupt Mask Register (TIMSK). The TIMSK register is used by all three timer channels. Timer 1 uses the OCIE1A/1B bits, the Timer/Counter 1 Overflow Interrupt Enable (TOIE1) bit, and the Timer/Counter 1 Input Capture Interrupt Enable (TIC1E1) bit. When the OCIE1A/B bit and the I-bit in the Status Register are both set to 1, the Timer/Counter 1 Compare Match interrupt is enabled. When the TOIE1 bit and the I-bit in the Status Register are both set to 1, the Timer/Counter 1 Overflow interrupt is enabled. When the TIC1E1 bit and the I-bit in the Status Register are both set to 1, the Timer/Counter 1 Input Capture interrupt is enabled.

5.7.1.7 Timer/Counter Interrupt Flag Register (TIFR). The TIMSK register is used by all three timer channels. Timer 1 uses the OCF1A/B, which sets for an output compare A/B match. Timer 1 also uses the Timer/Counter 1 Overflow Flag (TOV1), which sets when Timer/Counter 1 overflows. Timer Channel 1 also uses the Timer/Counter 1 Input Capture Flag (ICF1), which sets for an input capture event.

5.8 TIMER 2

Timer 2 is another 8-bit timer channel similar to Timer 0. The Timer 2 channel block diagram is provided in Figure 5.13. Its registers are summarized in Figure 5.14.

5.8.1 Timer/Counter Control Register 2

The TCCR2 register bits are used to

- select the operational mode of Timer 2 using the Waveform Mode Generation (WGM2[1:0]) bits,

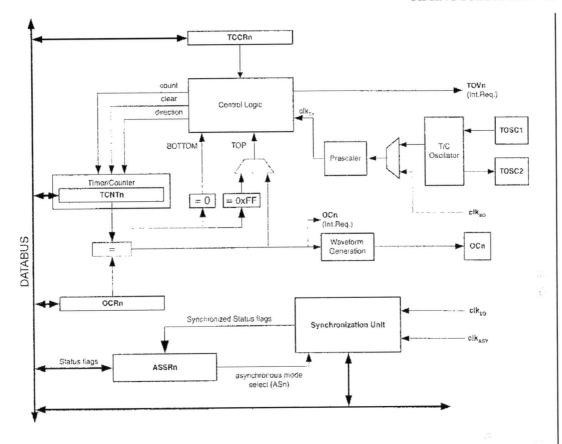

FIGURE 5.13: Timer 2 block diagram. Figure used with permission of Atmel.

- determine the operation of the timer within a specific mode with the Compare Match Output Mode (COM2[1:0]) bits, and
- select the source of the Timer 2 clock using CS2[2:0] bits.

The bit settings for the TCCR2 register are summarized in Figure 5.15.

5.8.2 Timer/Counter Register (TCNT2)
The TCNT2 is the 8-bit counter for Timer 2.

5.8.3 Output Compare Register (OCR2)
The OCR2 register holds a user-defined 8-bit value that is continuously compared with the TCNT2 register.

Timer/Counter Control Register (TCCR2)

FOC2	WGM20	COM21	COM20	WGM21	CS22	CS21	CS20

7   0

Timer/Counter Register (TCNT2)

7   0

Output Compare Register (OCR2)

7   0

Timer/Counter Interrupt Mask Register (TIMSK)

OCIE2	TOIE2	TICIE1	OCIE1A	OCIE1B	TOIE1	OCIE0	TOIE0

7   0

Timer/Counter Interrupt Flag REgister (TIFR)

OCF2	TOV2	ICF1	OCF1A	OCF1B	TOV1	OCF0	TOV0

7   0

FIGURE 5.14: Timer 2 registers.

5.8.4 Timer/Counter Interrupt Mask Register (TIMSK)

The TIMSK register is used by all three timer channels. Timer 2 uses the OCIE2 bit and the Timer/Counter 2 Overflow Interrupt Enable (TOIE2) bit. When the OCIE2 bit and the I-bit in the Status Register are both set to 1, the Timer/Counter 2 Compare Match interrupt is enabled. When the TOIE2 bit and the I-bit in the Status Register are both set to 1, the Timer/Counter 2 Overflow interrupt is enabled.

5.8.5 Timer/Counter Interrupt Flag Register

The TIMSK register is used by all three timer channels. Timer 2 uses the OCF2, which sets for an output compare match. Timer 2 also uses the Timer/Counter 2 Overflow Flag (TOV2), which sets when Timer/Counter 2 overflows.

5.9 PROGRAMMING THE TIMER SYSTEM

In this section, we provide several representative examples of using the timer system for various applications. We will provide examples of using the timer system to generate a prescribed delay and a PWM signal and to capture an input event.

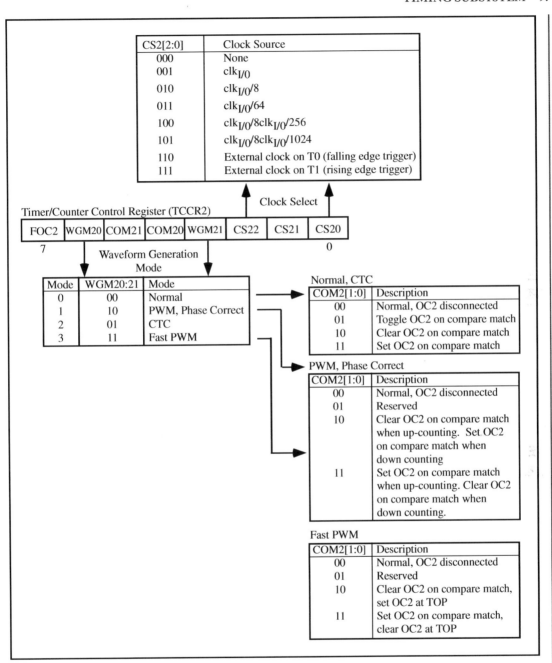

FIGURE 5.15: TCCR2 bit settings.

5.9.1 Precision Delay

In this example, we program the ATmega16 to provide a delay of some number of 6.55-ms interrupts. The Timer 0 overflow is configured to occur every 6.55 ms. The overflow flag is used as a "clock tick" to generate a precision delay. To create the delay the microcontroller is placed in a while loop waiting for the prescribed number of Timer 0 overflows to occur.

```
//Function prototypes
void delay(unsigned int number_of_6_55ms_interrupts);
void init_timer0_ovf_interrupt(void);
void timer0_interrupt_isr(void);

                                       //interrupt handler
                                       //definition
#pragma interrupt_handler timer0_interrupt_isr:10
                                       //door profile data


//*************************************************************
//int_timer0_ovf_interrupt(): The Timer0 overflow interrupt is
//being employed as a time base for a master timer for this
//project. The ceramic resonator operating at 10 MHz is divided
//by 256. The 8-bit Timer0 register (TCNT0) overflows every 256
//counts or every 6.55 ms.
//*************************************************************

void init_timer0_ovf_interrupt(void)
{
TCCR0 = 0x04; //divide timer0 timebase by 256, overflow occurs
              //every 6.55ms
TIMSK = 0x01; //enable timer0 overflow interrupt
asm("SEI");   //enable global interrupt
}

//*************************************************************
//*************************************************************
//timer0_interrupt_isr:
//Note: Timer overflow 0 is cleared by hardware when executing the
```

```c
//corresponding interrupt handling vector.
//***************************************************************
void timer0_interrupt_isr(void)
{
input_delay++;      //input delay processing
}


//***************************************************************
//***************************************************************
//delay(unsigned int num_of_6_55ms_interrupts): this generic delay
//function provides the specified delay as the number
//of 6.55 ms "clock ticks" from the Timer0 interrupt.
//Note: this function is only valid when using a 10 MHz crystal or
//        ceramic resonator
//***************************************************************

void delay(unsigned int number_of_6_55ms_interrupts)
{
TCNT0 = 0x00;                       //reset timer0
input_delay = 0;
while(input_delay <= number_of_6_55ms_interrupts)
   {
   ;
   }
}


//***************************************************************
```

5.9.2 Pulse Width Modulation

The function provided below is used to configure output compare Channel B to generate a PWM signal. An analog voltage provided to ADC channel 3 is used to set the desired duty cycle from 50% to 100%. Note how the PWM ramps up from 0 to the desired speed.


```c
//Function Prototypes
void PWM(unsigned int PWM_incr)
```

```c
{
unsigned int    Open_Speed_int;
float           Open_Speed_float;
int             gate_position_int;

PWM_duty_cycle = 0;
InitADC();                                  //Initialize ADC
                                            //Read "Open Speed" volt
                                            //setting PA3
Open_Speed_int = ReadADC(0x03);             //Open Speed Setting
                                            //unsigned int
                                            //Convert to max duty
                                            //cycle setting 0 VDC =
                                            //50% = 127, 5 VDC =
                                            //100% = 255
Open_Speed_float = ((float)(Open_Speed_int)/(float)(0x0400));
                                            //convert volt to
                                            //PWM constant 127-255
Open_Speed_int = (unsigned int)((Open_Speed_float * 127) + 128.0);
                                            //Configure PWM clock
TCCR1A = 0xA1;                              //freq = resonator/510
                                            //     = 10 MHz/510
                                            //freq = 19.607 kHz
TCCR1B = 0x01;                              //no clock source
                                            //division
                                            //Initiate PWM duty cycle
                                            //variables
PWM_duty_cycle = 0;

OCR1BH = 0x00;
OCR1BL = (unsigned char)(PWM_duty_cycle);//set PWM duty cycle CH
                                            //B to 0%
                                            //Ramp up to Open Speed
                                            //in 1.6s
OCR1BL = (unsigned char)(PWM_duty_cycle);//set PWM duty cycle CH
                                            //B
```

```
while (PWM_duty_cycle < Open_Speed_int)
  {
  if(PWM_duty_cycle < Open_Speed_int)     //increment duty cycle
    PWM_duty_cycle=PWM_duty_cycle + PWM_open_incr;


    OCR1BL = (unsigned char)(PWM_duty_cycle);//set PWM duty cycle
                                             //CH B

  }

//Gate continues to open at specified upper speed (PA3)
:
:
:
//*************************************************************
```

5.9.3 Input Capture Mode

This example was developed by Julie Sandberg, BSEE, and Kari Fuller, BSEE, at the University of Wyoming as part of their senior design project. In this example, the input capture channel is being used to monitor the heart rate (typically 50–120 beats per minute) of a patient. The microcontroller is set to operate at an internal clock frequency of 1 MHz.

```
//*************************************************************
//*************************************************************
//initialize_ICP_interrupt: Initialize Timer/Counter 1
//for input capture
//*************************************************************

void initialize_ICP_interrupt(void)
{
TIMSK=0x20;                       //Allows input capture
                                  //interrupts
SFIOR=0x04;                       //Internal pull-ups disabled
TCCR1A=0x00;                      //No output comp or waveform
                                  //generation mode
TCCR1B=0x45;                      //Capture on rising edge,
```

```c
                                        //clock prescalar=1024
TCNT1H=0x00;                            //Initially clear timer/
                                        //counter 1
TCNT1L=0x00;
asm("SEI");                             //enable global interrupts
}


//*****************************************************************
void Input_Capture_ISR(void)



{
if(first_edge==0)
   {
   ICR1L=0x00;                          //clear ICR1 and TCNT1
                                        //on first edge
   ICR1H=0x00;
   TCNT1L=0x00;
   TCNT1H=0x00;
   first_edge=1;
   }

else
   {
   ICR1L=TCNT1L;                        //capture time from TCNT1
   ICR1H=TCNT1H;
   TCNT1L=0x00;
   TCNT1H=0x00;
   first_edge=0;
   }

heart_rate();                           //calculate the heart rate
TIFR=0x20;                              //clear the input capture flag
asm("RETI");                            //resets the I flag to allow
                                        //global interrupts

}
```

```
/***************************************************************/
void heart_rate(void)
{
if(first_edge==0)
   {
   time_pulses_low = ICR1L;           //Read 8 low bits first
   time_pulses_high = ((unsigned int)(ICR1H << 8));
   time_pulses = time_pulses_low | time_pulses_high;
   if(time_pulses!=0)                 //1 counter increment = 1.024 ms
      {                               //divide by 977 to get seconds/
                                      //pulse
      HR=60/(time_pulses/977);        //(secs/min)/(secs/beat) =bpm
      }
   else
      {
      HR=0;
      }
   }
else
   {
   HR=0;
   }
}
/***************************************************************/
```

5.10 SUMMARY

In this chapter, we considered a microcontroller timer system, associated terminology for timer-related topics, discussed typical functions of a timer subsystem, studied timer hardware operations, and considered some applications where the timer subsystem of a microcontroller can be used. We then took a detailed look at the timer subsystem aboard the ATmega16 and reviewed the features, operation, registers, and programming of the three timer channels.

5.11 REFERENCES AND FURTHER READING

1. S Barrett and D Pack, *Microcontrollers Fundamentals for Engineers and Scientists*, Morgan & Claypool, San Rafael, CA, 2006. doi:10.2200/S00025ED1V01Y200605DCS001

2. *Atmel 8-bit AVR Microcontroller with 16K Bytes In-System Programmable Flash, ATmega16, ATmega16L*, data sheet: 2466L-AVR-06/05, Atmel, San Jose, CA.
3. T Morton, *Embedded Microcontrollers*, Prentice-Hall, Upper Saddle River, NJ, 2001.
4. K Short, *Embedded Microprocessor Systems Design: An Introduction Using the INTEL 80C188EB*, Prentice-Hall, Upper Saddle River, NJ, 1998.
5. F Driscoll, R Coughlin, and R Villanucci, *Data Acquisition and Process Control with the M68HC11 Microcontroller*, 2nd ed, Prentice-Hall, Upper Saddle River, NJ, 2000.

5.12 CHAPTER PROBLEMS

1. Question: Given an 8-bit free-running counter and a system clock rate of 24 MHz, find the time required for the counter to count from 0 to its maximum value.
2. Question: If we desire to generate periodic signals with periods ranging from 125 ns to 500 μs, what is the minimum frequency of the system clock?
3. Question: Describe how you can compute the period of an incoming signal with varying duty cycles.
4. Question: Describe how one can generate an aperiodic pulse with a pulse width of 2 min.
5. Question: Program the output compare system of the ATmega16 to generate a 1-kHz signal with a 10% duty cycle.
6. Question: Design a microcontroller system to control a sprinkler controller that performs the following tasks. We assume that your microcontroller runs with a 10-MHz clock and it has a 16-bit free-running counter. The sprinkler controller system controls two different zones by turning sprinklers within each zone on and off. To turn on the sprinklers of a zone, the controller needs to receive a 152.589-Hz PWM signal from your microcontroller. To turn off the sprinklers of the same zone, the controller needs to receive the PWM signal with a different duty cycle.
    a. Your microcontroller needs to provide the PWM signal with 10% duty cycle for 10 ms to turn on the sprinklers in zone 1.
    b. After 15 min, your microcontroller must send the PWM signal with 15% duty cycle for 10 ms to turn off the sprinklers in zone 1.
    c. After 15 min, your microcontroller must send the PWM signal with 20% duty cycle for 10 ms to turn on the sprinklers in zone 2.
    d. After 15 min, your microcontroller must send the PWM signal with 25% duty cycle for 10 ms to turn off the sprinklers in zone 2.

CHAPTER 6

Atmel AVR Operating Parameters and Interfacing

Objectives: After reading this chapter, the reader should be able to

- describe the voltage and current parameters for the Atmel AVR HC CMOS-type microcontroller,
- apply the voltage and current parameters toward properly interfacing I/O devices to the Atmel AVR microcontroller,
- interface a wide variety of I/O devices to the Atmel AVR microcontroller,
- describe the special concerns that must be followed when the Atmel AVR microcontroller is used to interface to a high-power DC or AC device,
- discuss the requirement for an optical-based interface,
- describe how to control the speed and direction of a DC motor, and
- describe how to control several types of AC loads.

Our first textbook for M&C, *Microcontrollers Fundamentals for Engineers and Scientists*, contained a chapter entitled "Operating Parameters and Interfacing" [1]. With M&C's permission, we have repeated this chapter here for your convenience. However, we have personalized the information provided to the Atmel AVR line of microcontrollers. We have also expanded the coverage of the chapter to include interface techniques for a number of additional I/O devices.

In this chapter, we introduce you to the extremely important concepts of the operating envelope for a microcontroller. We begin by reviewing the voltage and current electrical parameters for the HC CMOS-based Atmel AVR line of microcontrollers. We then show how to apply this information to properly interface I/O devices to the ATmega16 microcontroller. We then discuss the special considerations for controlling a high-power DC or AC load such as a motor and introduce the concept of an optical interface. Throughout the chapter, we provide a number of detailed examples.

6.1 OPERATING PARAMETERS

The microcontroller is an electronic device that has precisely defined operating conditions. As long as the microcontroller is used within its defined operating parameter limits, it should continue to operate correctly. However, if the allowable conditions are violated, spurious results may result.

Any time a device is connected to a microcontroller, careful interface analysis must be performed. Most microcontrollers are members of the "HC," or high-speed CMOS family of chips. As long as all components in a system are also of the "HC" family, as is the case for the Atmel AVR line of microcontrollers, electrical interface issues are minimal. If the microcontroller is connected to some component not in the HC family, electrical interface analysis must be completed. Manufacturers readily provide the electrical characteristic data necessary to complete this analysis in their support documentation.

To perform the interface analysis, there are eight different electrical specifications required for electrical interface analysis. The electrical parameters are

- V_{OH}: the lowest guaranteed output voltage for a logic high,
- V_{OH}: the highest guaranteed output voltage for a logic low,
- I_{OH}: the output current for a V_{OH} logic high,
- I_{OH}: the output current for a V_{OH} logic low,
- V_{IH}: the lowest input voltage guaranteed to be recognized as a logic high,
- V_{IL}: the highest input voltage guaranteed to be recognized as a logic low,
- I_{IH}: the input current for a V_{IH} logic high, and
- I_{IL}: the input current for a V_{IL} logic low.

These electrical characteristics are required for both the microcontroller and the external components. Typical values for a microcontroller in the HC CMOS family assuming $V_{DD} = 5.0$ V and $V_{SS} = 0$ V are provided below. The minus sign on several of the currents indicates a current flow out of the device. A positive current indicates current flow into the device.

- $V_{OH} = 4.2$ V,
- $V_{OL} = 0.4$ V,
- $I_{OH} = -0.8$ mA,
- $I_{OL} = 1.6$ mA,
- $V_{IH} = 3.5$ V,
- $V_{IL} = 1.0$ V,
- $I_{IH} = 10\ \mu A$, and
- $I_{IL} = -10\ \mu A$.

It is important to realize that these are static values taken under very specific operating conditions. If external circuitry is connected such that the microcontroller acts as a current source (current leaving microcontroller) or current sink (current entering microcontroller), the voltage parameters listed above will also be affected.

In the current source case, an output voltage V_{OH} is provided at the output pin of the microcontroller when the load connected to this pin draws a current of I_{OH}. If a load draws more current from the output pin than the I_{OH} specification, the value of V_{OH} is reduced. If the load current becomes too high, the value of V_{OH} falls below the value of V_{IH} for the subsequent logic circuit stage and not be recognized as an acceptable logic high signal. When this situation occurs, erratic and unpredictable circuit behavior results.

In the sink case, an output voltage V_{OL} is provided at the output pin of the microcontroller when the load connected to this pin delivers a current of I_{OL} to this logic pin. If a load delivers more current to the output pin of the microcontroller than the I_{OL} specification, the value of V_{OL} increases. If the load current becomes too high, the value of V_{OL} rises above the value of V_{IL} for the subsequent logic circuit stage and not be recognized as an acceptable logic low signal. As before, when this situation occurs, erratic and unpredictable circuit behavior results.

For convenience, this information is illustrated in Figure 6.1. In Figure 6.1(a), we have provided an illustration of the direction of current flow from the HC device and a comparison of voltage levels. As a reminder, current flowing out of a device is considered a negative current (source case), whereas current flowing into the device is considered positive current (sink case). The magnitude of the voltage and current for HC CMOS devices are shown in Figure 6.1(b). As more current is sunk or sourced from a microcontroller pin, the voltage will be pulled up or pulled down, respectively, as shown in Figure 6.1(c). If I/O devices are improperly interfaced to the microcontroller, these loading conditions may become excessive, and voltages will not be properly interpreted as the correct logic levels.

You must also ensure that total current limits for an entire microcontroller port and overall bulk port specifications are complied with. For planning purposes, the sum of current sourced or sunk from a port should not exceed 100 mA. Furthermore, the sum of currents for all ports should not exceed 200 mA. As before, if these guidelines are not complied with, erratic microcontroller behavior may result.

The procedures presented in the following sections when followed carefully will ensure the microcontroller will operate within its designed envelope. The remainder of the chapter is divided into input device interface analysis followed by output device interface analysis.

6.2 INPUT DEVICES

In this section, we discuss how to properly interface input devices to a microcontroller. We will start with the most basic input component, a simple on/off switch.

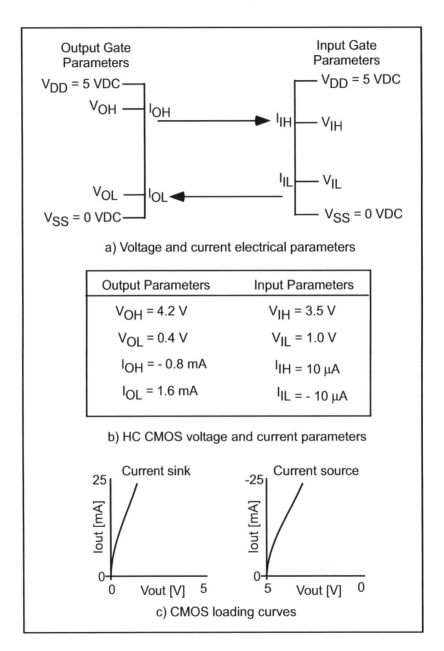

FIGURE 6.1: Electrical voltage and current parameters: (a) voltage and current electrical parameters, (b) HC CMOS voltage and current parameters, and (c) CMOS loading curves.

6.2.1 Switches

Switches come in a variety of types. As a system designer, it is up to you to choose the appropriate switch for a specific application. Switch varieties commonly used in microcontroller applications are illustrated in Figure 6.2(a). Here is a brief summary of the different types:

- **Slide switch:** A slide switch has two different positions: on and off. The switch is manually moved to one position or the other. For microcontroller applications, slide switches are

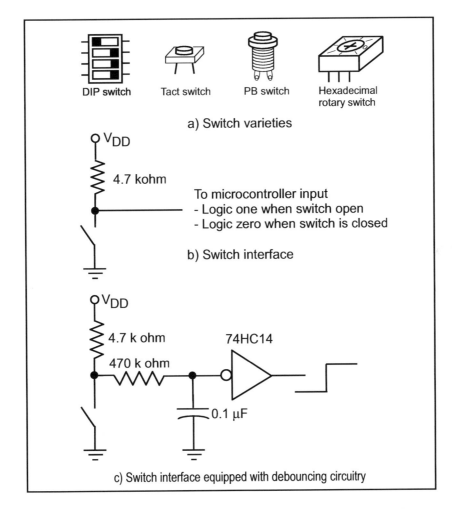

FIGURE 6.2: Switch interface: (a) switch varieties, (b) switch interface, and (c) switch interface equipped with debouncing circuitry.

available that fit in the profile of a common integrated circuit size DIP. A bank of four or eight DIP switches in a single package is commonly available.

- **Momentary contact push-button switch:** A momentary contact push-button switch comes in two varieties: normally closed (NC) and normally open (NO). A NO switch, as its name implies, does not normally provide an electrical connection between its contacts. When the push-button portion of the switch is depressed the connection between the two switch contacts is made. The connection is held as long as the switch is depressed. When the switch is released, the connection is opened. The converse is true for an NC switch. For microcontroller applications, push-button switches are available in a small tact type switch configuration.

- **Push on/push off switches:** These type of switches are also available in an NO or NC configuration. For the NO configuration, the switch is depressed to make connection between the two switch contacts. The push button must be depressed again to release the connection.

- **Hexadecimal rotary switches:** Small profile rotary switches are available for microcontroller applications. These switches commonly have 16 rotary switch positions. As the switch is rotated to each position a unique 4-bit binary code is provided at the switch contacts.

A common switch interface is shown in Figure 6.2(b). This interface allows a logic 1 or 0 to be properly introduced to a microcontroller input port pin. The basic interface consists of the switch in series with a current limiting resistor. The node between the switch and the resistor is provided to the microcontroller input pin. In the configuration shown, the resistor pulls the microcontroller input up to the supply voltage V_{DD}. When the switch is closed, the node is grounded, and a logic 0 is provided to the microcontroller input pin. To reverse the logic of the switch configuration, the position of the resistor and the switch is simply reversed.

6.2.2 Switch Debouncing

Mechanical switches do not make a clean transition from one position (on) to another (off). When a switch is moved from one position to another, it makes and breaks contact multiple times. This activity may go on for tens of milliseconds. A microcontroller is relatively fast as compared with the action of the switch. Therefore, the microcontroller is able to recognize each switch bounce as a separate and erroneous transition.

To correct the switch bounce phenomena, additional external hardware components may be used or software techniques may be employed. A hardware debounce circuit is illustrated in Figure 6.2(c). The node between the switch and the limiting resistor of the basic switch circuit is fed to a low pass filter (LPF) formed by the 470-kΩ resistor and the capacitor. The LPF prevents abrupt changes (bounces) in the input signal from the microcontroller. The LPF is followed by a 74HC14 Schmitt Trigger, which is simply an inverter equipped with hysteresis. This further limits the switch bouncing.

Switches may also be debounced using software techniques. This is accomplished by inserting a 30- to 50-ms lockout delay in the function responding to port pin changes. The delay prevents the microcontroller from responding to the multiple switch transitions related to bouncing.

You must carefully analyze a given design to determine if hardware or software switch debouncing techniques will be used. It is important to remember that all switches exhibit bounce phenomena and therefore must be debounced.

6.2.3 Keypads

A keypad is simply an extension of the simple switch configuration. A typical keypad configuration and interface are shown in Figure 6.3. As you can see, the keypad is simply multiple switches in the same package. A hexadecimal keypad is provided in the figure. A single row of keypad switches is asserted by the microcontroller, and then the host keypad port is immediately read. If a switch has been depressed, the keypad pin corresponding to the column the switch is in will also be asserted. The combination of a row and a column assertion can be decoded to determine which key has been pressed as illustrated in the table. Keypad rows are continually asserted one after the other in sequence. Because the keypad is a collection of switches, debounce techniques must also be employed.

The keypad may be used to introduce user requests to a microcontroller. A standard keypad with alphanumeric characters may be used to provide alphanumeric values to the microcontroller such as providing your personal identification number (PIN) for a financial transaction. However, some keypads are equipped with removable switch covers such that any activity can be associated with a key press.

6.2.4 Sensors

A microcontroller is typically used in control applications where data are collected, assimilated, and processed by the host algorithm and a control decision and accompanying signals are provided by the microcontroller. Input data for the microcontroller are collected by a complement of input sensors. These sensors may be digital or analog in nature.

6.2.4.1 Digital Sensors. Digital sensors provide a series of digital logic pulses with sensor data encoded. The sensor data may be encoded in any of the parameters associated with the digital pulse

112 ATMEL AVR MICROCONTROLLER PRIMER: PROGRAMMING AND INTERFACING

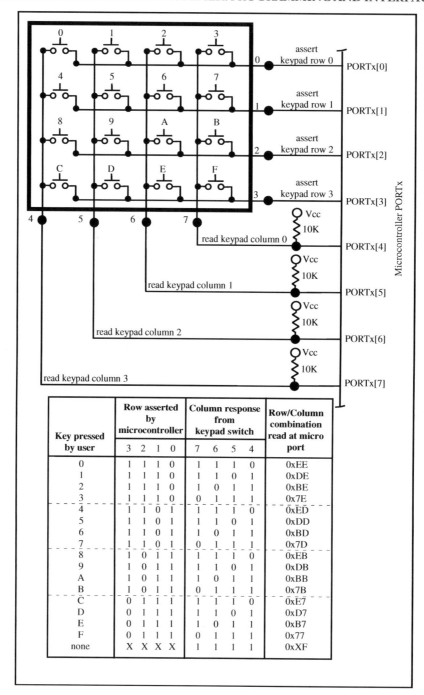

FIGURE 6.3: Keypad interface.

train such as duty cycle, frequency, period, or pulse rate. The input portion of the timing system may be configured to measure these parameters.

An example of a digital sensor is the optical encoder. An optical encoder consists of a small plastic transparent disk with opaque lines etched into the disk surface. A stationary optical emitter and detector source are placed on either side of the disk. As the disk rotates, the opaque lines break the continuity between the optical source and detector. The signal from the optical detector is monitored to determine disk rotation as shown in Figure 6.4.

Optical encoders are available in a variety of types, depending on the information desired. There are two major types of optical encoders: incremental and absolute encoders. An absolute encoder is used when it is required to retain position information when power is lost. For example, if you were using an optical encoder in a security gate control system, an absolute encoder would be

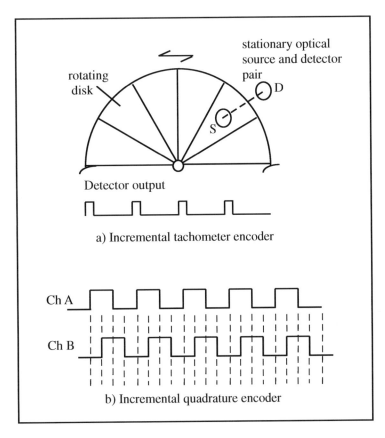

FIGURE 6.4: Optical encoder: (a) incremental tachometer encoder and (b) incremental quadrature encoder.

used to monitor the gate position. An incremental encoder is used in applications where a velocity or a velocity and direction information is required.

The incremental encoder types may be further subdivided into tachometers and quadrature encoders. An incremental tachometer encoder consists of a single track of etched opaque lines as shown in Figure 6.4(a). It is used when the velocity of a rotating device is required. To calculate velocity, the number of detector pulses is counted in a fixed amount of time. Because the number of pulses per encoder revolution is known, velocity may be calculated.

The quadrature encoder contains two tracks shifted in relationship to one another by 90°. This allows the calculation of both velocity and direction. To determine direction one would monitor the phase relationship between Channel A and Channel B as shown in Figure 6.4(b). The absolute encoder is equipped with multiple data tracks to determine the precise location of the encoder disk (SICK Stegmann [2]).

6.2.4.2 Analog Sensors.
Analog sensors provide a DC voltage that is proportional to the physical parameter being measured. As discussed in the ADC chapter, the analog signal may be first preprocessed by external analog hardware such that it falls within the voltage references of the conversion subsystem. The analog voltage is then converted to a corresponding binary representation.

An example of an analog sensor is the flex sensor shown in Figure 6.5(a). The flex sensor provides a change in resistance for a change in sensor flexure. At 0° flex, the sensor provides 10 kΩ of resistance. For 90° flex, the sensor provides 30–40 kΩ of resistance. Because the microcontroller cannot measure resistance directly, the change in flex sensor resistance must be converted to a change in a DC voltage. This is accomplished using the voltage divider network shown in Figure 6.5(c). For increased flex, the DC voltage will increase. The voltage can be measured using the ATmega16's ADC subsystem.

The flex sensor may be used in applications such as virtual reality data gloves, robotic sensors, biometric sensors, and in science and engineering experiments (Images Company [3]). One of the coauthors used the circuit provided in Figure 6.5 to help a colleague in zoology monitor the movement of a newt salamander during a scientific experiment.

6.3   OUTPUT DEVICES

As previously mentioned, an external device should not be connected to a microcontroller without first performing careful interface analysis to ensure the voltage, current, and timing requirements of the microcontroller and the external device are met. In this section, we describe interface considerations for a wide variety of external devices. We begin with the interface for a single LED.

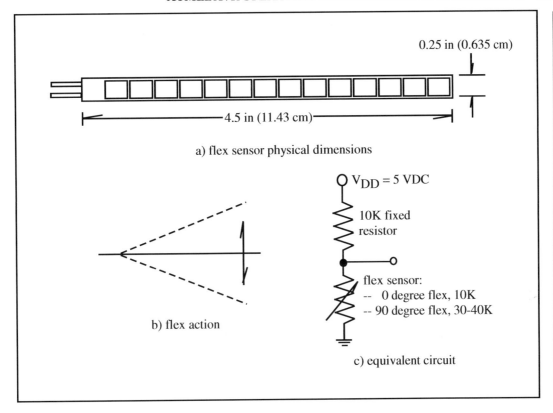

FIGURE 6.5: Flex sensor: (a) flex sensor's physical dimensions, (b) flex action, and (c) equivalent circuit.

6.3.1 Light-Emitting Diodes

A LED is typically used as a logic indicator to inform the presence of a logic 1 or a logic 0 at a specific pin of a microcontroller. An LED has two leads: the anode or positive lead and the cathode or negative lead. To properly bias an LED, the anode lead must be biased at a level approximately 1.7 to 2.2 V higher than the cathode lead. This specification is known as the forward voltage (V_f) of the LED. The LED current must also be limited to a safe level known as the forward current (I_f). The diode voltage and current specifications are usually provided by the manufacturer.

An example of an LED biasing circuit is provided in Figure 6.6(a). A logic 1 is provided by the microcontroller to the input of the inverter. The inverter provides a logic 0 at its output, which provides a virtual ground at the cathode of the LED. Therefore, the proper voltage biasing for the LED is provided. The resistor (R) limits the current through the LED. A proper resistor

116 ATMEL AVR MICROCONTROLLER PRIMER: PROGRAMMING AND INTERFACING

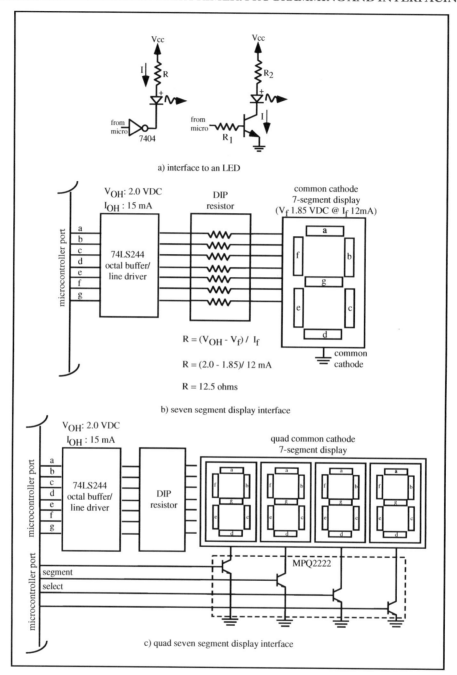

FIGURE 6.6: LED display devices: (a) interface to an LED, (b) seven-segment display interface, and (c) quad seven-segment display interface.

value can be calculated using $R = (V_{DD} - V_{DIODE})/I_{DIODE}$. It is important to note that a 7404 inverter must be used because its capability to safely sink 16 mA of current. Alternately, an NPN transistor such as a 2N2222 (PN2222 or MPQ2222) may be used in place of the inverter as shown in the figure.

6.3.2 Seven-Segment LED Displays

To display numeric data, seven-segment LED displays are available as shown in Figure 6.6(b). Different numerals can be displayed by asserting the proper LED segments. For example, to display the number 5, segments a, c, d, f, and g would be illuminated. Seven-segment displays are available in common cathode (CC) and common anode (CA) configurations. As the CC designation implies, all seven individual LED cathodes on the display are tied together.

The microcontroller is not capable of driving the LED segments directly. As shown in Figure 6.6(b), an interface circuit is required. We use a 74LS244 octal buffer/driver circuit to boost the current available for the LED. The LS244 is capable of providing 15 mA per segment (I_{OH}) at 2.0 VDC (V_{OH}). A limiting resistor is required for each segment to limit the current to a safe value for the LED. Conveniently, resistors are available in DIP packages of eight for this type of application.

Seven-segment displays are available in multicharacter panels. In this case, separate microcontroller ports are not used to provide data to each seven-segment character. Instead, a single port is used to provide character data. A portion of another port is used to sequence through each of the characters as shown in Figure 6.6(c). An NPN (for a CC display) transistor is connected to the common cathode connection of each individual character. As the base contact of each transistor is sequentially asserted the specific character is illuminated. If the microcontroller sequences through the display characters at a rate greater than 30 Hz, the display will have steady illumination.

6.3.3 Tristate LED Indicator

The tristate LED indicator introduced in Chapter 1 is shown in Figure 6.7. It is used to provide the status of an entire microcontroller port. The indicator bank consists of eight green and eight red LEDs. When an individual port pin is logic high, the green LED is illuminated. When logic low the red LED is illuminated. If the port pin is at a tristate high-impedance state, no LED is illuminated.

The NPN/PNP transistor pair at the bottom of the figure provides a 2.5-VDC voltage reference for the LEDs. When a specific port pin is logic high (5.0 VDC), the green LED will be forward biased because its anode will be at a higher potential than its cathode. The 47-Ω resistor limits current to a safe value for the LED. Conversely, when a specific port pin is at a logic low (0 VDC), the red LED will be forward biased and illuminate. For clarity, the red and green LEDs are shown as being separate devices. LEDs are available that have both LEDs in the same device.

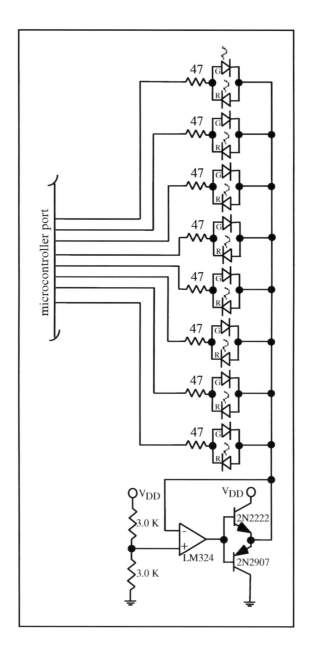

FIGURE 6.7: Tristate LED display.

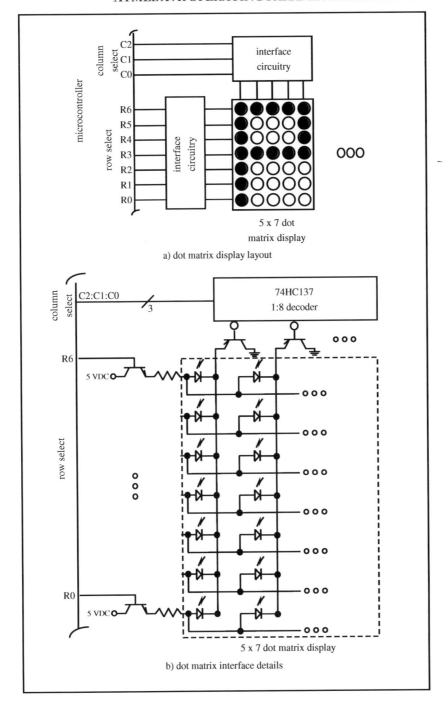

FIGURE 6.8: Dot matrix display: (a) dot matrix display and (b) dot matrix interface details.

6.3.4 Dot Matrix Display

The dot matrix display consists of a large number of LEDs configured in a single package. A typical 5×7 LED arrangement is a matrix of five columns of LEDs with seven LEDs per row as shown in Figure 6.8. Display data for a single matrix column [R6-R0] is provided by the microcontroller. That specific row is then asserted by the microcontroller using the column select lines (C2–C0). The entire display is sequentially built up a column at a time. If the microcontroller sequences through each column fast enough (greater than 30 Hz), the matrix display appears to be stationary to a human viewer.

In Figure 6.8(a), we have provided the basic configuration for the dot matrix display for a single-display device. However, this basic idea can be expanded in both dimensions to provide a multicharacter, multiline display. A larger display does not require a significant number of microcontroller pins for the interface. The dot matrix display may be used to display alphanumeric data as well as graphics data. In Figure 6.8(b), we have provided additional detail of the interface circuit.

6.3.5 Liquid Crystal Display

An LCD is an output device to display text information as shown in Figure 6.9. LCDs come in a wide variety of configurations including multicharacter, multiline format. A 16×2 LCD format is common. That is, it has the capability of displaying two lines of 16 characters each. The characters are sent to the LCD via ASCII format a single character at a time. For a parallel-configured LCD,

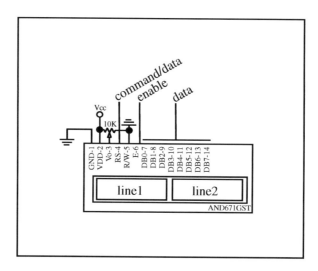

FIGURE 6.9: LCD display.

ATMEL AVR OPERATING PARAMETERS AND INTERFACING 121

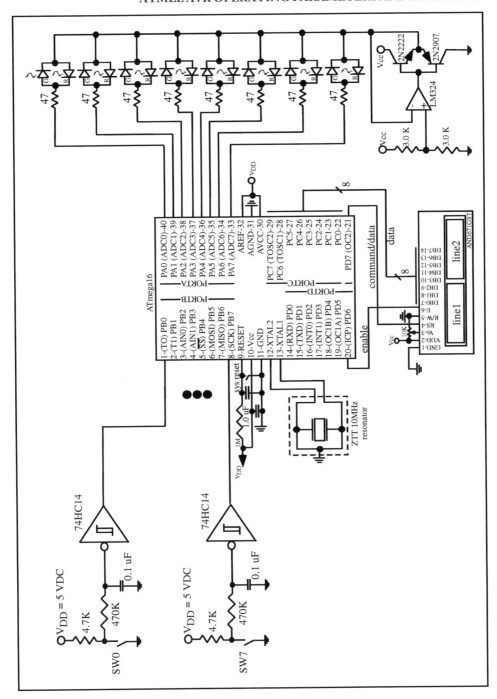

FIGURE 6.10: Hardware testbench equipped with an LCD.

an 8-bit data path and two lines are required between the microcontroller and the LCD. A small microcontroller mounted to the back panel of the LCD translates the ASCII data characters and control signals to properly display the characters. LCDs are configured for either parallel or serial data transmission format. In the example provided, we use a parallel-configured display. In Figure 6.10, we have included the LCD in the Testbench hardware configuration.

Some sample C code is provided below to send data and control signals to an LCD. In this specific example, an AND671GST 1 × 16 character LCD was connected to the Atmel ATmega16 microcontroller [4]. One 8-bit port and two extra control lines are required to connect the microcontroller to the LCD. Note: The initialization sequence for the LCD is specified within the manufacturer's technical data.

```
//***********************************************************
//LCD_Init: initialization for an LCD connected in the following
//manner:
//LCD: AND671GST 1x16 character display
//LCD configured as two 8 character lines in a 1x16 array
//LCD data bus (pin 14-pin7)    ATMEL ATmega16: PORTC
//LCD RS (pin 4)    ATMEL ATmega16: PORTD[7]
//LCD E  (pin 6)    ATMEL ATmega16: PORTD[6]
//***********************************************************

void LCD_Init(void)
{
delay_5ms();
delay_5ms();
delay_5ms();
                              // output command string to
                              // initialize LCD
putcommand(0x38);             //function set 8-bit
delay_5ms();
putcommand(0x38);             //function set 8-bit
putcommand(0x38);             //function set 8-bit
putcommand(0x38);             //one line, 5x7 char
putcommand(0x0C);             //display on
putcommand(0x01);             //display clear-1.64 ms
putcommand(0x06);             //entry mode set
```

```c
putcommand(0x00);              //clear display, cursor at home
putcommand(0x00);              //clear display, cursor at home
}


//***************************************************************
//putchar:prints specified ASCII character to LCD
//***************************************************************

void putchar(unsigned char c)
{
DDRC = 0xff;                   //set PORTC as output
DDRD = DDRD|0xC0;              //make PORTD[7:6] output
PORTC = c;
PORTD = PORTD|0x80;            //RS=1
PORTD = PORTD|0x40;            //E=1
PORTD = PORTD&0xbf;            //E=0
delay_5ms();
}


//***************************************************************
//performs specified LCD related command
//***************************************************************

void putcommand(unsigned char d)
{
DDRC = 0xff;                   //set PORTC as output
DDRD = DDRD|0xC0;              //make PORTD[7:6] output
PORTD = PORTD&0x7f;            //RS=0
PORTC = d;
PORTD = PORTD|0x40;            //E=1
PORTD = PORTD&0xbf;            //E=0
delay_5ms();
}

//***************************************************************
```

6.3.6 High-Power DC Devices

A number of direct current devices may be controlled with an electronic switching device such as a MOSFET. Specifically, an N-channel enhancement MOSFET (metal oxide semiconductor field effect transistor) may be used to switch a high-current load on and off (such as a motor) using a low-current control signal from a microcontroller as shown in Figure 6.11(a). The low-current control signal from the microcontroller is connected to the gate of the MOSFET. The MOSFET switches the high-current load on and off consistent with the control signal. The high-current load is connected between the load supply and the MOSFET drain. It is important to note that the load supply voltage and the microcontroller supply voltage do not have to be at the same value. When the control signal on the MOSFET gate is logic high, the load current flows from drain to source. When the control signal applied to the gate is logic low, no load current flows. Thus, the high-power load is turned on and off by the low-power control signal from the microcontroller.

Often the MOSFET is used to control a high-power motor load. A motor is a notorious source of noise. To isolate the microcontroller from the motor noise, an optical isolator may be used as an interface as shown in Figure 6.11(b). The link between the control signal from the microcontroller to the high-power load is via an optical link contained within an SSR. The SSR is properly biased using techniques previously discussed.

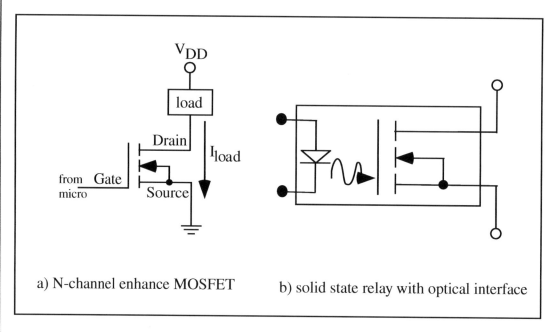

FIGURE 6.11: MOSFET circuits: (a) N-channel enhance MOSFET and (b) solid-state relay (SSR) with optical interface.

6.4 DC MOTOR SPEED AND DIRECTION CONTROL

Often, a microcontroller is used to control a high-power motor load. To properly interface the motor to the microcontroller, we must be familiar with the different types of motor technologies. Motor types are illustrated in Figure 6.12.

- **DC motor:** A DC motor has a positive and negative terminal. When a DC power supply of suitable current rating is applied to the motor, it will rotate. If the polarity of the supply is switched with reference to the motor terminals, the motor will rotate in the opposite direction. The speed of the motor is roughly proportional to the applied voltage up to the rated voltage of the motor.

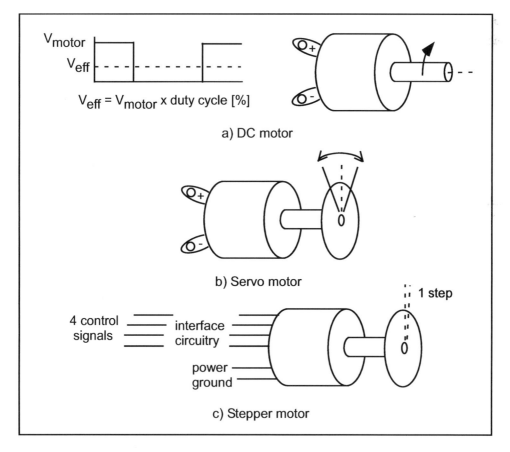

FIGURE 6.12: Motor types: (a) DC, (b) servo, and (c) stepper.

- **Servo motor:** A servo motor provides a precision angular rotation for an applied PWM duty cycle. As the duty cycle of the applied signal is varied, the angular displacement of the motor also varies. This type of motor is used to change mechanical positions such as the steering angle of a wheel.
- **Stepper motor:** A stepper motor, as its name implies, provides an incremental step change in rotation (typically 2.5° per step) for a step change in control signal sequence. The motor is typically controlled by a two- or four-wire interface. For the four-wire stepper motor, the microcontroller provides a 4-bit control sequence to rotate the motor clockwise. To turn the motor counterclockwise, the control sequence is reversed. The low-power control signals are interfaced to the motor via MOSFETs or power transistors to provide for the proper voltage and current requirements of the pulse sequence.

6.4.1 DC Motor Operating Parameters

Space does not allow a full discussion of all motor types. We will concentrate on the DC motor. As previously mentioned, the motor speed may be varied by changing the applied voltage. This is difficult to do with a digital control signal. However, PWM control signal techniques discussed earlier may be combined with a MOSFET interface to precisely control the motor speed. The duty cycle of the PWM signal will also be the percentage of the motor supply voltage applied to the motor and hence the percentage of rated full speed at which the motor will rotate. The interface circuit to accomplish this type of control is shown in Figure 6.13. Various portions of this interface circuit have been previously discussed. The resistor R_G, typically 10 kΩ, is provided to discharge the MOSFET gate when no voltage is applied to the gate. For an inductive load, a reversed biased protection diode must be provided across the load. The interface circuit shown allows the motor to rotate in a given direction. As previously mentioned, to rotate the motor in the opposite direction, the motor polarity must be reversed. This may be accomplished with a high-power switching network called an H-bridge specifically designed for this purpose. Reference Pack and Barrett [5,6] for more information on this topic.

6.4.2 AC Devices

In a similar manner, a high-power AC load may be switched on and off using a low-power control signal from the microcontroller. In this case, an SSR is used as the switching device. SSRs are available to switch a high-power DC or AC load (Crydom) [7]. For example, the Crydom 558-CX240D5R is a printed circuit board mounted, air-cooled, single-pole, single-throw (SPST), NO SSR. It requires a DC control voltage of 3–15 VDC at 15 mA. However, this small

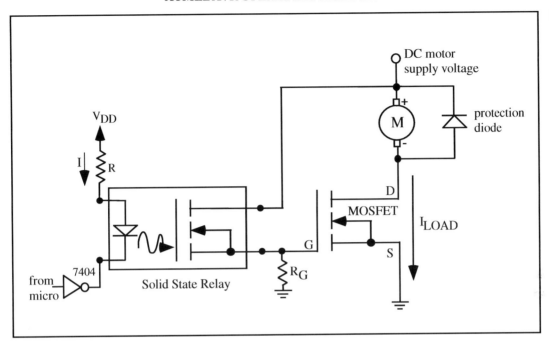

FIGURE 6.13: DC motor interface.

microcontroller-compatible DC control signal is used to switch 12- to 280-VAC loads rated from 0.06 to 5 A (Crydom).

To vary the direction of an AC motor, you must use a bidirectional AC motor. A bidirectional motor is equipped with three terminals: common, clockwise, and counterclockwise. To turn the motor clockwise, an AC source is applied to the common and clockwise connections. In like manner, to turn the motor counterclockwise, an AC source is applied to the common and counterclockwise connections. This may be accomplished using two of the Crydom SSRs.

6.5 APPLICATION: FLIGHT SIMULATOR PANEL

We close the chapter with an extended example of developing a flight simulator panel. This panel was actually designed and fabricated for a middle school to allow students to react to space mission-like status while using a program that allowed them to travel about the planets.

An Atmel ATmega8 microcontroller was used because its capabilities best fit the requirements for the project. We will retain the use of the ATmega8 in the example to illustrate the ease and transferring information from one microcontroller to another in the Atmel AVR line.

The panel face is shown in Figure 6.14. It consists of a joystick that is connected to a host computer for the flight simulator software. Below the joystick is a two-line LCD equipped with a backlight LED (Hantronix HDM16216L-7, Jameco# 658988). Below the LCD is a buzzer to

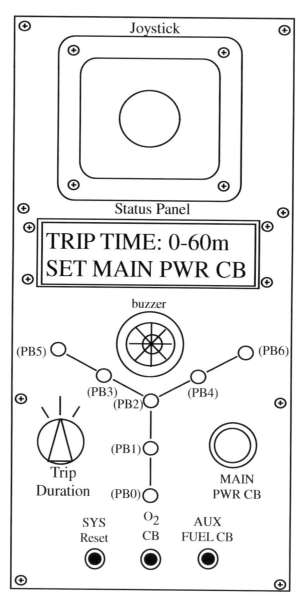

FIGURE 6.14: Flight simulator panel.

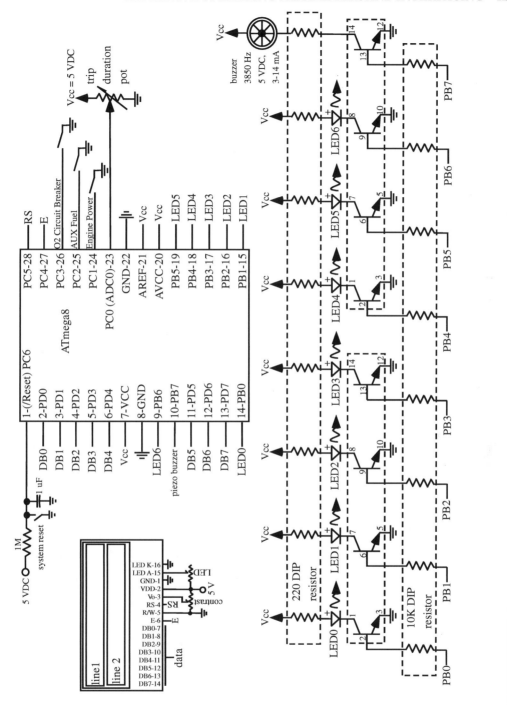

FIGURE 6.15: Interface diagram for the flight simulator panel.

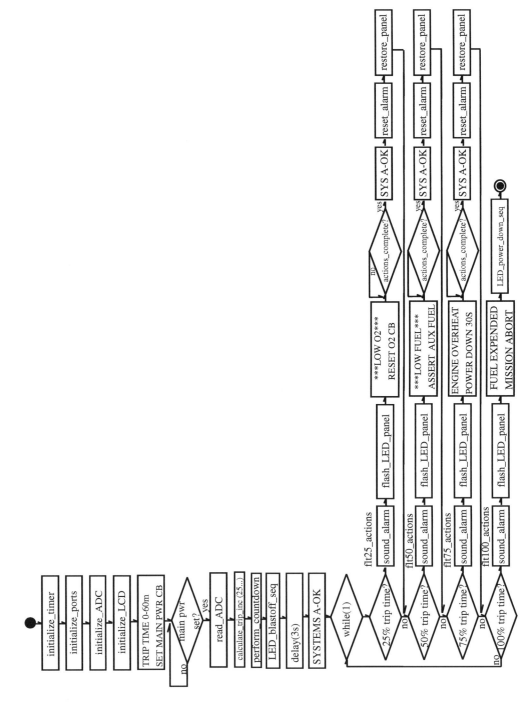

FIGURE 6.16: Software flow for the flight simulator panel.

alert students to changing status. There is also a complement of other status indicators. From left to right is the Trip Duration potentiometer. At the beginning of the flight episode, students are prompted to set this from 0 to 60 min to communicate to the microcontroller the length of the flight episode. These data are used to calculate different flight increments. There are also a series of simulated circuit breakers: system reset (SYS Reset), oxygen (O_2 CB), auxiliary fuel (AUX FUEL CB), and the main power circuit breakers (MAIN PWR CB). These are not actual circuit breakers but NO SPST momentary push-button switches that allow the students to interact with the microcontroller. There are also a series of LEDs that form a Y pattern on the panel face. They are also used to indicate status changes.

To interface the flight simulator panel to the microcontroller, a number of different techniques previously discussed in the book were employed. The interface diagram is shown in Figure 6.15. Pin 1 is a reset for the microcontroller. When the switch is depressed, pin 1 is taken low and resets the microcontroller. Port D of the microcontroller (pins 2–6, 11–13) forms the data connection for the LCD. Pin 9 is used to turn the buzzer on and off. This pin is routed through a transistor interface described earlier in the text. Port B[5:0] (pins 14–19) are used to control the LEDs on the front panel. Each individual LED is also supported with a transistor interface circuit. Conveniently, these small NPN signal transistors come in fours to a 14-pin DIP package (MPQ2222). Port C [0] (pin 23) is used as an analog input pin. It is connected to the trip duration potentiometer. Port C pins [3:1] (pins 24–26) are used to connect the NO SPST tact switches to the microcontroller. Port C pins [4:5] (pins 27 and 28) are used for the LCD control signals.

The software flowchart is shown in Figure 6.16. After startup the students are prompted via the LCD display to set the trip duration and then press the main power circuit breaker. This potentiometer setting is then used to calculate four different trip increments. Countdown followed by blastoff then commences. At four different trip time increments, students are presented with status that they must respond to. Real clock time is kept using the TCNT0 timer overflow configured as a 65.5 ms "clock tick." The overall time base for the microcontroller was its internal 1 MHz clock that may be selected during programming with the STK500.

Provided below is the code listing for the flight simulator panel.

```
//****************************************************************
//file name: flight_sim.c
//author: Steve Barrett, Ph.D., P.E.
//created: August 1, 2007
//last revised: August 19, 2007
//function: Controls Flight Simulator Control Panel for Larimer
//          County School District #1
```

```
//
//Compiler settings:
//   Paths: c:\icc\include\
//   Lib:   c:\icc\lib
//   Output Directory: c:\output\
//Compiler Options:
//   Accept Extensions
//   Optimizxations: Default
//   AVR Studio Version: Studio 4.06
//   ATmega8
//   printf: small
//   Return Stack: 16
//
//AVR Program Tab Settings
//Program:
//   Select device: ATmega8
//   Programming mode: ISP, Erase Device Before Programming,
//                     Verify Device After Programming
//   Flash: Input HEX file, Browse and find machine code file:
//          flight_sim.hex
//   EEPROM: Input HEX file, Browse and find machine code file:
//          flight_sim.EEP
//Fuses: Set the following fuses
//   Boot flash section size = 128
//   Brown out detection at Vcc = 4.0V
//   Internal Oscillator: 1 MHz
//Lock bits:
//   Mode 1
//   Application Protection Mode 1
//   Boot loader Protection Mode 1
//Advanced:  N/A
//Board: N/A
//Auto:
//   Erase Device
//   Program Flash
//   Verify Flash
```

```
//   Program Fuses
//   Verify Fuses
//   Read Fuses
//
//ATMEL AVR ATmega8
//Chip Port Function I/O Source/Dest Asserted Notes
//*****************************************************************
//Pin 1: /Reset
//Pin 2: PD0 to DB0 LCD
//Pin 3: PD1 to DB1 LCD
//Pin 4: PD2 to DB2 LCD
//Pin 5: PD3 to DB3 LCD
//Pin 6: PD4 to DB4 LCD
//Pin 7: Vcc
//Pin 8: Gnd
//Pin 9: PB6 to LED6
//Pin 10: PB7 to piezo buzzer
//Pin 11: PD5 to DB6 LCD
//Pin 12: PD6 to DB6 LCD
//Pin 13: PD7 to DB7 LCD
//Pin 14: PB0 to LED0
//Pin 15: PB1 to LED1
//Pin 16: PB2 to LED2
//Pin 17: PB3 to LED3
//Pin 18: PB4 to LED4
//Pin 19: PB5 to LED5
//Pin 20: AVCC to Vcc
//Pin 21: AREF to Vcc
//Pin 22 Gnd
//Pin 23 ADC0 to trip duration potentiometer
//Pin 24 PC1 Engine Power Switch
//Pin 25 PC2 AUX Fuel circuit breaker
//Pin 26 PC3 O2 circuit breaker
//Pin 27 PC4 to LCD Enable (E)
//Pin 28 PC5 to LCD RS
//
```

```c
//include files************************************************

//ATMEL register definitions for ATmega8
#include<iom8v.h>


//function prototypes*******************************************
void delay(unsigned int number_of_65_5ms_interrupts);
void init_timer0_ovf_interrupt(void);
void InitADC(void);                           //initialize ADC
void initialize_ports(void);                  //initializes ports
void power_on_reset(void);                    //returns system to
                                              //startup state
unsigned int  ReadADC(unsigned char chan);    //read value from ADC
                                              //results
void clear_LCD(void);                         //clears LCD display
void LCD_Init(void);                          //initialize AND671GST
                                              //LCD
void putchar(unsigned char c);                //send character to LCD
void putcommand(unsigned char c);             //send command to LCD
unsigned int  ReadADC(unsigned char chan);    //read value from ADC
                                              //results
void timer0_interrupt_isr(void);
void flt25_actions(void);
void flt50_actions(void);
void flt75_actions(void);
void flt100_actions(void);
void sound_alarm(void);
void turn_off_LEDs(void);
void reset_alarm(void);
void restore_panel(void);
void LED_blastoff_sequence(void);
void LED_power_down_sequence(void);
void monitor_main_power_CB(void);
void monitor_O2_CB_reset(void);
void monitor_aux_fuel_CB(void);
void perform_countdown(void);
```

```c
void print_LOWO2(void);
void print_LOW_FUEL(void);
void print_fuel_expended(void);
void print_OVERHEAT(void);
void print_trip_dur(void);
void flash_LED_panel(void);
void clear_LCD(void);
void calculate_trip_int(void);
void systems_A_OK(void);

//program constants
#define TRUE    1
#define FALSE   0
#define OPEN    1
#define CLOSE   0
#define YES     1
#define NO      0
#define SAFE    1
#define UNSAFE  0
#define ON      1
#define OFF     0

//interrupt handler definition
#pragma interrupt_handler timer0_interrupt_isr:10

//main program*****************************************

//global variables
unsigned int flt_25, flt_50, flt_75, flt_100;
unsigned int action25_done=NO, action50_done=NO;
unsigned int action75_done=NO, action100_done=NO;
unsigned int achieved25=NO, achieved50=NO;
unsigned int achieved75=NO, achieved100=NO;
unsigned int flt_timer=0;
unsigned int trip_duration_volt;
unsigned char PORTC_pullup_mask = 0x0e;
```

```c
unsigned int flash_timer;
unsigned int PORTB_LEDs;
unsigned int flash_panel=NO;
unsigned int delay_timer;
unsigned int troubleshooting = 1;
void convert_display_voltage_LCD(int trip_duration_volt);
void convert_int_to_string_display_LCD(unsigned int
total_integer_value);

void main(void)
{
init_timer0_ovf_interrupt();         //initialize Timer0 to
                                     //serve as elapsed
initialize_ports();                  //initialize ports
InitADC();                           //initialize ADC
LCD_Init();                          //initialize LCD
print_trip_dur();                    //prompt user to enter trip
                                     //duration
monitor_main_power_CB();
clear_LCD();

trip_duration_volt = ReadADC(0x00);  //Read trip duration ADC0
if(troubleshooting)
   {
   convert_display_voltage_LCD(trip_duration_volt);//display
   voltage LCD delay(46);
   }
calculate_trip_int();
if(troubleshooting)
   {
   convert_int_to_string_display_LCD(flt_25);
   delay(46);
   }
perform_countdown();
LED_blastoff_sequence();
sound_alarm();
```

```c
delay(46);
reset_alarm();
systems_A_OK();

while(1)
  {
  if(flt_timer > flt_25)   achieved25  = YES;
  if(flt_timer > flt_50)   achieved50  = YES;
  if(flt_timer > flt_75)   achieved75  = YES;
  if(flt_timer > flt_100)  achieved100 = YES;

    if((achieved25==YES)&&(action25_done==NO))     //25% flight
                                                   //    complete

      {
      flt25_actions();
      action25_done=YES;
  systems_A_OK();
      }

    if((achieved50==YES)&&(action50_done==NO))     //50% flight
                                                   //    complete

      {
      flt50_actions();
      action50_done=YES;
  systems_A_OK();
      }

    if((achieved75==YES)&&(action75_done==NO))     //75% flight
                                                   //    complete

      {
      flt75_actions();
      action75_done=YES;
  systems_A_OK();
      }
    if((achieved100==YES)&&(action100_done==NO))   //100% flight
                                                   //    complete
```

```c
         {
         flt100_actions();
         action100_done=YES;
         }
      }//end while
}//end main


//function definitions*****************************************
//***************************************************************
//initialize_ports: provides initial configuration for I/O ports
//
//Note: when the RSTDISBL fuse is unprogrammed, the RESET
//      circuitry is connected to the pin, and the pin cannot be
//      used as an I/O pin.
//***************************************************************

void initialize_ports(void)
{
DDRB = 0xff;                   //PORTB[7:0] as output
PORTB= 0x00;                   //initialize low
DDRC = 0xb0;                   //set PORTC as output   OROO_IIII
                               //                      1011_0000
PORTC= PORTC_pullup_mask;      //initialize pullups PORTC[3:1]
DDRD = 0xff;                   //set PORTD as output
PORTD =0x00;                   //initialize low
}


//***************************************************************
//delay(unsigned int num_of_65_5ms_interrupts): this generic
//delay function
//provides the specified delay as the number of 65.5 ms
//"clock ticks" from the
//Timer0 interrupt.
//Note: this function is only valid when using a 1 MHz crystal or
//ceramic resonator
//***************************************************************
```

```c
void delay(unsigned int number_of_65_5ms_interrupts)
{
TCNT0 = 0x00;                               //reset timer0
delay_timer = 0;
while(delay_timer <= number_of_65_5ms_interrupts)
   {
   ;
   }
}

//****************************************************************
//InitADC: initialize ADC converter
//****************************************************************
void InitADC( void)
{
ADMUX = 0;                                  //Select channel 0
ADCSRA = 0xC3;                              //Enable ADC & start 1st
                                            //dummy conversion
                                            //Set ADC module prescalar
                                            //to 8
    //critical for accurate ADC results
while (!(ADCSRA & 0x10));                   //Check if conversation is
                                            //ready
ADCSRA |= 0x10;                             //Clear conv rdy flag-set
                                            //the bit
}

//****************************************************************
//ReadADC: read analog voltage from ADC - the desired channel for
//conversion
//is passed in as an unsigned character variable.  The result is
//returned
//as a left justified, 10 bit binary result.  The ADC prescalar
//must be set to 8 to slow down the ADC clock at higher external
//clock frequencies (10 MHz) to obtain accurate results.
//****************************************************************
```

```c
unsigned int ReadADC(unsigned char channel)
{
unsigned int binary_weighted_voltage, binary_weighted_voltage_low;
unsigned int binary_weighted_voltage_high; //weighted binary
                                           //voltage

ADMUX = channel;                           //Select channel
ADCSRA |= 0x43;                            //Start conversion
                                           //Set ADC module prescalar
                                           //to 8
  //critical for accurate ADC results
while (!(ADCSRA & 0x10));                  //Check if converstion is
                                           //ready
ADCSRA |= 0x10;                            //Clear Conv rdy flag - set
                                           //the bit
binary_weighted_voltage_low = ADCL;        //Read 8 low bits first
                                           //(important)
                                           //Read 2 high bits,multiply
                                           //by 256
binary_weighted_voltage_high = ((unsigned int)(ADCH << 8));
binary_weighted_voltage=binary_weighted_voltage_low|
binary_weighted_voltage_high;
return binary_weighted_voltage;            //ADCH:ADCL
}

//*****************************************************************
//int_timer0_ovf_interrupt(): The Timer0 overflow interrupt is
//being employed as a time base for a master timer for this
//project. The internal time base is set to operate at 1 MHz and
//then is divided by 256.  The 8-bit Timer0 register (TCNT0)
//overflows every 256 counts or every 65.5 ms.
//*****************************************************************

void init_timer0_ovf_interrupt(void)
{
TCCR0 = 0x04; //divide timer0 timebase by 256, overflow occurs
```

```c
                    //every 65.5ms
TIMSK = 0x01;  //enable timer0 overflow interrupt
asm("SEI");    //enable global interrupt
}


//***************************************************************
//LCD_Init: initialization for an LCD connected in the following
//manner:
//LCD: AND671GST 1x16 character display
//LCD configured as two 8 character lines in a 1x16 array
//LCD data bus (pin 14-pin7) ATMEL 8: PORTD
//LCD RS (pin 28)   ATMEL 8: PORTC[5]
//LCD E  (pin 27)   ATMEL 8: PORTC[4]
//***************************************************************

void LCD_Init(void)
{
delay(1);
delay(1);
delay(1);
                   // output command string to initialize LCD
putcommand(0x38);  //function set 8-bit
delay(1);
putcommand(0x38);  //function set 8-bit
putcommand(0x38);  //function set 8-bit
putcommand(0x38);  //one line, 5x7 char
putcommand(0x0C);  //display on
putcommand(0x01);  //display clear-1.64 ms
putcommand(0x06);  //entry mode set
putcommand(0x00);  //clear display, cursor at home
putcommand(0x00);  //clear display, cursor at home
}

//***************************************************************
//putchar:prints specified ASCII character to LCD
//***************************************************************
```

```c
void putchar(unsigned char c)
{
DDRD  = 0xff;                                   //set PORTD as output
DDRC  = DDRC|0x30;                              //make PORTC[5:4] output
PORTD = c;
PORTC = (PORTC|0x20)|PORTC_pullup_mask;    //RS=1
PORTC = (PORTC|0x10)|PORTC_pullup_mask;;   //E=1
PORTC = (PORTC&0xef)|PORTC_pullup_mask;;   //E=0
delay(1);
}


//***************************************************************
//putcommand: performs specified LCD related command
//***************************************************************

void putcommand(unsigned char d)
{
DDRD  = 0xff;                                   //set PORTD as output


DDRC  = DDRC|0xC0;                              //make PORTA[5:4] output
PORTC = (PORTC&0xdf)|PORTC_pullup_mask;    //RS=0
PORTD = d;
PORTC = (PORTC|0x10)|PORTC_pullup_mask;    //E=1
PORTC = (PORTC&0xef)|PORTC_pullup_mask;    //E=0
delay(1);
}


//***************************************************************
//clear_LCD: clears LCD
//***************************************************************

void clear_LCD(void)
{
putcommand(0x01);
}
```

```c
//****************************************************************
//*void calculate_trip_int(void)
//****************************************************************

void calculate_trip_int(void)
{
unsigned int    trip_duration_sec;
unsigned int    trip_duration_int;

trip_duration_sec=(unsigned int)(((double)(trip_duration_volt)/
                1024.0)*60.0*60.0);
trip_duration_int = (unsigned int)((double)(trip_duration_sec)/
                 0.0655);
flt_25  = (unsigned int)((double)(trip_duration_int) * 0.25);
flt_50  = (unsigned int)((double)(trip_duration_int) * 0.50);
flt_75  = (unsigned int)((double)(trip_duration_int) * 0.75);
flt_100 = trip_duration_int;
}


//****************************************************************
//void timer0_interrupt_isr(void)
//****************************************************************

void timer0_interrupt_isr(void)
{
delay_timer++;
flt_timer++;                    //increment flight timer

if(flash_panel==YES)
   {
   if(flash_timer <= 8)
      {
      flash_timer++;
}
   else
      {
```

```c
   flash_timer = 0;
   if(PORTB_LEDs == OFF)
      {
      PORTB = 0xff;
      PORTB_LEDs = ON;
      }
   else
      {
      PORTB = 0x00;
      PORTB_LEDs = OFF;
      }
   }
    }
else
   {
   flash_timer = 0;
   }
}

//*************************************************************
//void flt25_actions(void)
//*************************************************************
void flt25_actions(void)
{
sound_alarm();
flash_LED_panel();
print_LOWO2();
monitor_O2_CB_reset();
reset_alarm();
restore_panel();
action25_done = YES;
}

//*************************************************************
//void flt50_actions(void)
//*************************************************************
```

```c
void flt50_actions(void)
{
sound_alarm();
flash_LED_panel();
print_LOW_FUEL();
monitor_aux_fuel_CB();
reset_alarm();
restore_panel();
action50_done = YES;
}

//*************************************************************
//void flt75_actions(void)
//*************************************************************

void flt75_actions(void)
{
sound_alarm();
flash_LED_panel();
print_OVERHEAT();
delay(458);                                 //delay 30s
monitor_main_power_CB();
reset_alarm();
restore_panel();
action75_done = YES;
}

//*************************************************************
//void flt100_actions(void)
//*************************************************************

void flt100_actions(void)
{
sound_alarm();
flash_LED_panel();
print_fuel_expended();
```

```c
turn_off_LEDs();
action100_done = YES;
}


//*************************************************************
//void sound_alarm(void)
//*************************************************************

void sound_alarm(void)
{
PORTB = PORTB | 0x80;
}


//*************************************************************
//void turn_off_LEDs(void)
//*************************************************************

void turn_off_LEDs(void)
{
PORTB = PORTB & 0x80;
}


//*************************************************************
//void reset_alarm(void)
//*************************************************************

void reset_alarm(void)
{
PORTB = PORTB & 0x7F;
}


//*************************************************************
//void restore_panel(void)
//*************************************************************

void restore_panel(void)
```

```c
{
flash_panel = NO;
PORTB = PORTB | 0x7F;
}

//***************************************************************
//void LED_blastoff_sequence(void)
//***************************************************************

void LED_blastoff_sequence(void)
{
PORTB = 0x00;           //0000_0000
delay(15);              //delay 1s
PORTB = 0x01;           //0000_0001
delay(15);              //delay 1s
PORTB = 0x03;           //0000_0011
delay(15);              //delay 1s
PORTB = 0x07;           //0000_0111
delay(15);              //delay 1s
PORTB = 0x1F;           //0001_1111
delay(15);              //delay 1s
PORTB = 0x7F;           //0111_1111
delay(15);              //delay 1s
}

//***************************************************************
//void LED_power_down_sequence(void)
//***************************************************************

void LED_power_down_sequence(void)
{
PORTB = 0x7F;           //0111_1111
delay(15);              //delay 1s
PORTB = 0x1F;           //0001_1111
delay(15);              //delay 1s
PORTB = 0x07;           //0000_0111
```

```c
  delay(15);              //delay 1s
  PORTB = 0x03;           //0000_0011
  delay(15);              //delay 1s
  PORTB = 0x01;           //0000_0001
  delay(15);              //delay 1s
  PORTB = 0x00;           //0000_0000
  delay(15);              //delay 1s
}

//*************************************************************
//void monitor_main_power_CB(void)
//*************************************************************

void monitor_main_power_CB(void)
{
  while((PINC & 0x02) == 0x02)
  {
  ;     //wait for PC1 to be exerted low
  }
}


//*************************************************************
//void monitor_O2_CB_reset(void)
//*************************************************************

void monitor_O2_CB_reset(void)
{
  while((PINC & 0x08) == 0x08)
    {
    ;     //wait for PC3 to be exerted low
    }
}


//*************************************************************
//void monitor_aux_fuel_CB(void)
//*************************************************************
```

```c
void monitor_aux_fuel_CB(void)
{
  while((PINC & 0x04) == 0x04)
    {
    ;    //wait for PC2 to be exerted low
    }
}

//**************************************************************
//void perform_countdown(void)
//**************************************************************

void perform_countdown(void)
{
clear_LCD();
putcommand(0x01);              //cursor home
putcommand(0x80);              //DD RAM location 1 - line 1
putchar('1'); putchar ('0');   //print 10
delay(15);                     //delay 1s

putcommand(0x01);              //cursor home
putcommand(0x80);              //DD RAM location 1 - line 1
putchar('9');                  //print 9
delay(15);                     //delay 1s

putcommand(0x01);              //cursor home
putcommand(0x80);              //DD RAM location 1 - line 1
putchar('8');                  //print 8
delay(15);                     //delay 1s

putcommand(0x01);              //cursor home
putcommand(0x80);              //DD RAM location 1 - line 1
putchar('7');                  //print 7
delay(15);                     //delay 1s

putcommand(0x01);              //cursor home
```

```c
putcommand(0x80);              //DD RAM location 1 - line 1
putchar('6');                  //print 6
delay(15);                     //delay 1s

putcommand(0x01);              //cursor home
putcommand(0x80);              //DD RAM location 1 - line 1
putchar('5');                  //print 5
delay(15);                     //delay 1s

putcommand(0x01);              //cursor home
putcommand(0x80);              //DD RAM location 1 - line 1
putchar('4');                  //print 4
delay(15);                     //delay 1s

putcommand(0x01);              //cursor home
putcommand(0x80);              //DD RAM location 1 - line 1
putchar('3');                  //print 3
delay(15);                     //delay 1s

putcommand(0x01);              //cursor home
putcommand(0x80);              //DD RAM location 1 - line 1
putchar('2');                  //print 2
delay(15);                     //delay 1s

putcommand(0x01);              //cursor home
putcommand(0x80);              //DD RAM location 1 - line 1
putchar('1');                  //print 1
delay(15);                     //delay 1s

putcommand(0x01);              //cursor home
putcommand(0x80);              //DD RAM location 1 - line 1
putchar('0');                  //print 0
delay(15);                     //delay 1s

//BLASTOFF!
putcommand(0x01);              //cursor home
```

```c
putcommand(0x80);              //DD RAM location 1 - line 1
putchar('B'); putchar('L'); putchar('A'); putchar('S'); putchar
                                                            ('T');
putchar('O'); putchar('F'); putchar('F'); putchar('!');
}

//*************************************************************
//void print_LOWO2(void)
//*************************************************************

void print_LOWO2(void)
{
clear_LCD();
putcommand(0x01);              //cursor home
putcommand(0x80);              //DD RAM location 1 - line 1
putchar('L'); putchar('O'); putchar('W'); putchar(' '); putchar
                                                            ('O');
putchar('2');

putcommand(0xC0);//DD RAM location 1 - line 2
putchar('R'); putchar('E'); putchar('S'); putchar('E'); putchar
                                                            ('T');
putchar(' '); putchar('O'); putchar('2'); putchar(' '); putchar
                                                            ('C');
putchar('B');
}

//*************************************************************
//void print_LOW_FUEL(void)
//*************************************************************

void print_LOW_FUEL(void)
{
clear_LCD();
putcommand(0x01);              //cursor home
putcommand(0x80);              //DD RAM location 1 - line 1
```

```c
putchar('L'); putchar('O'); putchar('W'); putchar(' '); putchar
                                                              ('F');
putchar('U'); putchar('E'); putchar('L');
putcommand(0xC0);//DD RAM location 1 - line 2
putchar('A'); putchar('S'); putchar('S'); putchar('E'); putchar
                                                              ('R');
putchar('T'); putchar(' '); putchar('A'); putchar('U'); putchar
                                                              ('X');
putchar('F'); putchar('U'); putchar('E'); putchar('L'); putchar
                                                              ('C');
putchar('B');
}

//*****************************************************************
//void print_fuel_expended(void)
//*****************************************************************
void print_fuel_expended(void)
{
clear_LCD();
putcommand(0x01);              //cursor home
putcommand(0x80);              //DD RAM location 1 - line 1
putchar('F'); putchar('U'); putchar('E'); putchar('L'); putchar
                                                              (' ');
putchar('E'); putchar('X'); putchar('P'); putchar('E'); putchar
                                                              ('N');
putchar('D'); putchar('E'); putchar('D');

putcommand(0xC0);//DD RAM location 1 - line 2
putchar('S'); putchar('H'); putchar('U'); putchar('T'); putchar
                                                              ('T');
putchar('I'); putchar('N'); putchar('G'); putchar(' '); putchar
                                                              ('D');
putchar('O'); putchar('W'); putchar('N'); putchar('.'); putchar
                                                              ('.');
putchar('.');
}
```

```c
//*************************************************************
//void print_trip_dur(void);
//*************************************************************

void print_trip_dur(void)
{
clear_LCD();
putcommand(0x01);              //cursor home
putcommand(0x80);              //DD RAM location 1 - line 1
putchar('T'); putchar('R'); putchar('I'); putchar('P');
putchar('T'); putchar('I'); putchar('M'); putchar('E'); putchar(':');
putchar('0'); putchar('-'); putchar('6'); putchar('0');

putcommand(0xC0);//DD RAM location 1 - line 2
putchar('S'); putchar('E'); putchar('T'); putchar(' '); putchar('M');
putchar('A'); putchar('I'); putchar('N'); putchar(' '); putchar('P');
putchar('W'); putchar('R'); putchar(' '); putchar('C'); putchar('B');
}

//*************************************************************
//void print_OVERHEAT(void)
//*************************************************************

void print_OVERHEAT(void)
{
clear_LCD();
putcommand(0x01);              //cursor home
putcommand(0x80);              //DD RAM location 1 - line 1
putchar('E'); putchar('N'); putchar('G'); putchar('I'); putchar('N');
putchar('E'); putchar(' '); putchar('O'); putchar('V'); putchar('E');
```

```c
putchar('R'); putchar('H'); putchar('E'); putchar('A'); putchar
                                                        ('T');

putcommand(0xC0);//DD RAM location 1 - line 2
putchar('R'); putchar('E'); putchar('S'); putchar('E'); putchar
                                                        ('T');
putchar(' '); putchar('E'); putchar('N'); putchar('G'); putchar
                                                        (' ');
putchar('C'); putchar('B'); putchar(' '); putchar('3'); putchar
                                                        ('0');
putchar('S');
}

//*************************************************************
//void systems_A_OK(void)
//*************************************************************

void systems_A_OK(void)
{
clear_LCD();
putcommand(0x01);              //cursor home
putcommand(0x80);              //DD RAM location 1 - line 1
putchar('S'); putchar('Y'); putchar('S'); putchar('T'); putchar
                                                        ('E');
putchar('M'); putchar('S'); putchar(' '); putchar('A'); putchar
                                                        ('-');
putchar('O'); putchar('K'); putchar('!'); putchar('!'); putchar
                                                        ('!');
}

//*************************************************************
//void flash_LED_panel(void)
//*************************************************************

void flash_LED_panel(void)
{
```

```c
    flash_panel = YES;
    flash_timer = 0;
    PORTB = 0x00;
    PORTB_LEDs = OFF;
}

//***************************************************************
//convert_display_voltage_LCD: converts binary weighted voltage
//to ASCII representation and prints result to LCD screen
//***************************************************************
void convert_display_voltage_LCD(int binary_voltage)
{
float   actual_voltage;                     //voltage between 0 and
                                            //5 V
int     all_integer_voltage;                //integer representation
                                            //of voltage
                                            //int representation of
                                            //voltage
int     hundreths_place, tens_place, ones_place;
                                            //char representation of
                                            //voltage
char    hundreths_place_char, tens_place_char, ones_place_char;

                                            // display analog voltage
                                            //on LCD
putcommand(0xC0);                           //LCD cursor to line 2
                                            //scale float voltage
                                            //0..5V
actual_voltage = ((float)(binary_voltage)/(float)(0x3FF))*5.0;
                                            //voltage represented 0
                                            //to 500
all_integer_voltage=actual_voltage * 100;//represent as all
                                            //integer
hundreths_place = all_integer_voltage/100;//isolate first digit
hundreths_place_char = (char)(hundreths_place + 48); //convert to
                                                     //ascii
```

```c
putchar(hundreths_place_char);          //display first digit
putchar('.');                           //print decimal point to
                                        //LCD
                                        //isolate tens place
xtens_place = (int)((all_integer_voltage - (hundreths_place*100))/
                                10);
tens_place_char=(char)(tens_place+48);  //convert to ASCII
putchar(tens_place_char);               //print to LCD
                                        //isolate ones place
ones_place = (int)((all_integer_voltage - (hundreths_place*100))
                                %10);
ones_place_char=(char)(ones_place+48);  //convert to ASCII
putchar(ones_place_char);               //print to LCD
putchar('V');                           //print unit V
}

//*************************************************************
//convert_int_to_string_display_LCD: converts 16 bit to unsigned
//integer
//values range from 0 to 65,535
//prints result to LCD screen
//*************************************************************

void convert_int_to_string_display_LCD(unsigned int total_
                                        integer_value)
{
int    ten_thousandths_place, thousandths_place;
int    hundreths_place, tens_place, ones_place;
char   ten_thousandths_place_char, thousandths_place_char;
char   hundreths_place_char, tens_place_char, ones_place_char;

putcommand(0xC0);                       //LCD cursor to line 2
                                        //10,000th place
ten_thousandths_place = total_integer_value/10000;
ten_thousandths_place_char = (char)(ten_thousandths_place+48);
putchar(ten_thousandths_place_char);
```

```c
                                            //1,000th place
thousandths_place = (int)((total_integer_value -
                    (ten_thousandths_place*10000))/1000);
thousandths_place_char = (char)(thousandths_place+48);
putchar(thousandths_place_char);
                                            //100th place
hundreths_place = (int)((total_integer_value -
                    (ten_thousandths_place*10000)-
(thousandths_place*1000))/100);
hundreths_place_char = (char)(hundreths_place + 48);
putchar(hundreths_place_char);
                                            //10th place
tens_place = (int)((total_integer_value -(ten_thousandths_place*
                                            10000)-
   (thousandths_place*1000)-(hundreths_place*100))/10);
tens_place_char=(char)(tens_place+48);    //convert to ASCII
putchar(tens_place_char);                 //print to LCD
                                            //isolate ones place
ones_place = (int)((total_integer_value -(ten_thousandths_place*
                                            10000)-
   (thousandths_place*1000)-(hundreths_place*100))%10);
ones_place_char=(char)(ones_place+48);    //convert to ASCII
putchar(ones_place_char);                 //print to LCD
}

//***********************************************************
//end of file: flight_sim.c
//***********************************************************
```

6.6 SUMMARY

In this chapter, we have discussed the voltage and current operating parameters for the Atmel HC CMOS-type microcontroller. We discussed how this information may be applied to properly design an interface for common I/O circuits. It must be emphasized that a properly designed interface allows the microcontroller to operate properly within its parameter envelope. If, because of a poor interface design, a microcontroller is used outside its prescribed operating parameter values, spurious and incorrect logic values will result. We provided interface information for a wide range of I/O devices. We also discussed the concept of interfacing a motor to a microcontroller using PWM techniques coupled with high-power MOSFET or SSR switching devices.

6.7 REFERENCES AND FURTHER READING

1. S Barrett and D Pack, *Microcontrollers Fundamentals for Engineers and Scientists*, Morgan & Claypool, San Rafael, CA, 2006. doi:10.2200/S00025ED1V01Y200605DCS001
2. SICK Stegmann Incorporated, Dayton, OH (www.stegmann.com).
3. Images Company, Staten Island, NY.
4. *Atmel 8-bit AVR Microcontroller with 16K Bytes In-System Programmable Flash, ATmega16, ATmega16L*, data sheet: 2466L-AVR-06/05, Atmel, San Jose, CA.
5. S Barrett and D Pack, *Embedded Systems Design with the 68HC12 and HCS12*, Prentice-Hall, Upper Saddle River, NJ, 2004.
6. D Pack and S Barrett, *68HC12 Microcontroller: Theory and Applications*, Prentice-Hall, Upper Saddle River, NJ, 2002.
7. Crydom Corporation, San Diego, CA (www.crydom.com).

6.8 CHAPTER PROBLEMS

1. Question: What will happen if a microcontroller is used outside its prescribed operating envelope?
2. Question: Discuss the difference between the terms *sink* and *source* as related to current loading of a microcontroller.
3. Question: Can an LED with a series limiting resistor be directly driven by the Atmel microcontroller? Explain.
4. Question: In your own words, provide a brief description of each of the microcontroller electrical parameters.
5. Question: What is switch bounce? Describe two techniques to minimize switch bounce.
6. Question: Describe a method of debouncing a keypad.
7. Question: What is the difference between an incremental encoder and an absolute encoder? Describe applications for each type.
8. Question: What must be the current rating of the 2N2222 and 2N2907 transistors used in the tristate LED circuit? Support your answer.
9. Question: Draw the circuit for a six-character, seven-segment display. Fully specify all components. Write a program to display "ATmega16."
10. Question: Repeat the question above for a dot matrix display.
11. Question: Repeat the question above for a LCD display.

APPENDIX A
ATmega16 Register Set

Register Summary

Address	Name	Bit 7	Bit 6	Bit 5	Bit 4	Bit 3	Bit 2	Bit 1	Bit 0	Page
$3F ($5F)	SREG	I	T	H	S	V	N	Z	C	7
$3E ($5E)	SPH	–	–	–	–	–	SP10	SP9	SP8	10
$3D ($5D)	SPL	SP7	SP6	SP5	SP4	SP3	SP2	SP1	SP0	10
$3C ($5C)	OCR0	Timer/Counter0 Output Compare Register								83
$3B ($5B)	GICR	INT1	INT0	INT2	–	–	–	IVSEL	IVCE	46, 67
$3A ($5A)	GIFR	INTF1	INTF0	INTF2	–	–	–	–	–	68
$39 ($59)	TIMSK	OCIE2	TOIE2	TICIE1	OCIE1A	OCIE1B	TOIE1	OCIE0	TOIE0	83, 114, 132
$38 ($58)	TIFR	OCF2	TOV2	ICF1	OCF1A	OCF1B	TOV1	OCF0	TOV0	84, 115, 132
$37 ($57)	SPMCR	SPMIE	RWWSB	–	RWWSRE	BLBSET	PGWRT	PGERS	SPMEN	250
$36 ($56)	TWCR	TWINT	TWEA	TWSTA	TWSTO	TWWC	TWEN	–	TWIE	178
$35 ($55)	MCUCR	SM2	SE	SM1	SM0	ISC11	ISC10	ISC01	ISC00	30, 66
$34 ($54)	MCUCSR	JTD	ISC2	–	JTRF	WDRF	BORF	EXTRF	PORF	39, 67, 229
$33 ($53)	TCCR0	FOC0	WGM00	COM01	COM00	WGM01	CS02	CS01	CS00	81
$32 ($52)	TCNT0	Timer/Counter0 (8 Bits)								83
$31 ($51)[1]	OSCCAL	Oscillator Calibration Register								28
	OCDR	On-Chip Debug Register								225
$30 ($50)	SFIOR	ADTS2	ADTS1	ADTS0	–	ACME	PUD	PSR2	PSR10	55,86,133,199,219
$2F ($4F)	TCCR1A	COM1A1	COM1A0	COM1B1	COM1B0	FOC1A	FOC1B	WGM11	WGM10	109
$2E ($4E)	TCCR1B	ICNC1	ICES1	–	WGM13	WGM12	CS12	CS11	CS10	112
$2D ($4D)	TCNT1H	Timer/Counter1 – Counter Register High Byte								113
$2C ($4C)	TCNT1L	Timer/Counter1 – Counter Register Low Byte								113
$2B ($4B)	OCR1AH	Timer/Counter1 – Output Compare Register A High Byte								113
$2A ($4A)	OCR1AL	Timer/Counter1 – Output Compare Register A Low Byte								113
$29 ($49)	OCR1BH	Timer/Counter1 – Output Compare Register B High Byte								113
$28 ($48)	OCR1BL	Timer/Counter1 – Output Compare Register B Low Byte								113
$27 ($47)	ICR1H	Timer/Counter1 – Input Capture Register High Byte								114
$26 ($46)	ICR1L	Timer/Counter1 – Input Capture Register Low Byte								114
$25 ($45)	TCCR2	FOC2	WGM20	COM21	COM20	WGM21	CS22	CS21	CS20	127
$24 ($44)	TCNT2	Timer/Counter2 (8 Bits)								129
$23 ($43)	OCR2	Timer/Counter2 Output Compare Register								129
$22 ($42)	ASSR	–	–	–	–	AS2	TCN2UB	OCR2UB	TCR2UB	130
$21 ($41)	WDTCR	–	–	–	WDTOE	WDE	WDP2	WDP1	WDP0	41
$20 ($40)[2]	UBRRH	URSEL	–	–	–	UBRR[11:8]				165
	UCSRC	URSEL	UMSEL	UPM1	UPM0	USBS	UCSZ1	UCSZ0	UCPOL	164
$1F ($3F)	EEARH	–	–	–	–	–	–	–	EEAR8	17
$1E ($3E)	EEARL	EEPROM Address Register Low Byte								17
$1D ($3D)	EEDR	EEPROM Data Register								17
$1C ($3C)	EECR	–	–	–	–	EERIE	EEMWE	EEWE	EERE	17
$1B ($3B)	PORTA	PORTA7	PORTA6	PORTA5	PORTA4	PORTA3	PORTA2	PORTA1	PORTA0	64
$1A ($3A)	DDRA	DDA7	DDA6	DDA5	DDA4	DDA3	DDA2	DDA1	DDA0	64
$19 ($39)	PINA	PINA7	PINA6	PINA5	PINA4	PINA3	PINA2	PINA1	PINA0	64
$18 ($38)	PORTB	PORTB7	PORTB6	PORTB5	PORTB4	PORTB3	PORTB2	PORTB1	PORTB0	64
$17 ($37)	DDRB	DDB7	DDB6	DDB5	DDB4	DDB3	DDB2	DDB1	DDB0	64
$16 ($36)	PINB	PINB7	PINB6	PINB5	PINB4	PINB3	PINB2	PINB1	PINB0	64
$15 ($35)	PORTC	PORTC7	PORTC6	PORTC5	PORTC4	PORTC3	PORTC2	PORTC1	PORTC0	65
$14 ($34)	DDRC	DDC7	DDC6	DDC5	DDC4	DDC3	DDC2	DDC1	DDC0	65
$13 ($33)	PINC	PINC7	PINC6	PINC5	PINC4	PINC3	PINC2	PINC1	PINC0	65
$12 ($32)	PORTD	PORTD7	PORTD6	PORTD5	PORTD4	PORTD3	PORTD2	PORTD1	PORTD0	65
$11 ($31)	DDRD	DDD7	DDD6	DDD5	DDD4	DDD3	DDD2	DDD1	DDD0	65
$10 ($30)	PIND	PIND7	PIND6	PIND5	PIND4	PIND3	PIND2	PIND1	PIND0	65
$0F ($2F)	SPDR	SPI Data Register								140
$0E ($2E)	SPSR	SPIF	WCOL	–	–	–	–	–	SPI2X	140
$0D ($2D)	SPCR	SPIE	SPE	DORD	MSTR	CPOL	CPHA	SPR1	SPR0	138
$0C ($2C)	UDR	USART I/O Data Register								161
$0B ($2B)	UCSRA	RXC	TXC	UDRE	FE	DOR	PE	U2X	MPCM	162
$0A ($2A)	UCSRB	RXCIE	TXCIE	UDRIE	RXEN	TXEN	UCSZ2	RXB8	TXB8	163
$09 ($29)	UBRRL	USART Baud Rate Register Low Byte								165
$08 ($28)	ACSR	ACD	ACBG	ACO	ACI	ACIE	ACIC	ACIS1	ACIS0	200
$07 ($27)	ADMUX	REFS1	REFS0	ADLAR	MUX4	MUX3	MUX2	MUX1	MUX0	215
$06 ($26)	ADCSRA	ADEN	ADSC	ADATE	ADIF	ADIE	ADPS2	ADPS1	ADPS0	217
$05 ($25)	ADCH	ADC Data Register High Byte								218
$04 ($24)	ADCL	ADC Data Register Low Byte								218
$03 ($23)	TWDR	Two-wire Serial Interface Data Register								180
$02 ($22)	TWAR	TWA6	TWA5	TWA4	TWA3	TWA2	TWA1	TWA0	TWGCE	180

FIGURE A.1: Atmel AVR ATmega16 Register Set. Figure used with permission of Atmel.

APPENDIX B
ATmega16 Header File

During C programming, the contents of a specific register may be referred to by name when an appropriate header file is included within your program. The header file provides the link between the register name used within a program and the hardware location of the register.

Provided below is the ATmega16 header file from the ICC AVR compiler. This header file was provided courtesy of ImageCraft.

```
#ifndef __iom16v_h
#define __iom16v_h

/* ATmega16 header file for
 * ImageCraft ICCAVR compiler
 */

/* not strictly sorted by address no more
 */

/* interrupt vector number definitions added
 */

/* last changed 2002/05/01
 */

/* 2 wire serial interface */
#define TWBR (*(volatile unsigned char *)0x20)
#define TWSR (*(volatile unsigned char *)0x21)
#define    TWPS1    1
#define    TWPS0    0
#define TWCR (*(volatile unsigned char *)0x56)
```

```
#define   TWINT    7
#define   TWEA     6
#define   TWSTA    5
#define   TWSTO    4
#define   TWWC     3
#define   TWEN     2
#define   TWIE     0
#define TWAR (*(volatile unsigned char *)0x22)
#define   TWGCE    0
#define TWDR (*(volatile unsigned char *)0x23)

/* ADC */
#define ADC   (*(volatile unsigned int *)0x24)
#define ADCL  (*(volatile unsigned char *)0x24)
#define ADCH  (*(volatile unsigned char *)0x25)
#define ADCSR (*(volatile unsigned char *)0x26)
#define ADCSRA (*(volatile unsigned char *)0x26)
#define   ADEN     7
#define   ADSC     6
#define   ADATE    5
#define   ADFR     5    /*This definition was in old ""datasheet*/
#define   ADIF     4
#define   ADIE     3
#define   ADPS2    2
#define   ADPS1    1
#define   ADPS0    0
#define ADMUX (*(volatile unsigned char *)0x27)
#define   REFS1    7
#define   REFS0    6
#define   ADLAR    5
#define   MUX4     4
#define   MUX3     3
#define   MUX2     2
#define   MUX1     1
#define   MUX0     0
```

```c
/* Analog Comparator */
#define ACSR (*(volatile unsigned char *)0x28)
#define  ACD    7
#define  ACBG   6
#define  ACO    5
#define  ACI    4
#define  ACIE   3
#define  ACIC   2
#define  ACIS1  1
#define  ACIS0  0

/* USART */
#define UBRRHI (*(volatile unsigned char *)0x40)
#define UBRRH (*(volatile unsigned char *)0x40)
#define  URSEL  7
#define UBRR (*(volatile unsigned char *)0x29)
#define UBRRL (*(volatile unsigned char *)0x29)
#define UCSRC (*(volatile unsigned char *)0x40)
#define  URSEL  7
#define  UMSEL  6
#define  UPM1   5
#define  UPM0   4
#define  USBS   3
#define  UCSZ1  2
#define  UCSZ0  1
#define  UCPOL  0
#define UCSRB (*(volatile unsigned char *)0x2A)
#define  RXCIE  7
#define  TXCIE  6
#define  UDRIE  5
#define  RXEN   4
#define  TXEN   3
#define  UCSZ2  2
#define  RXB8   1
#define  TXB8   0
#define UCSRA (*(volatile unsigned char *)0x2B)
```

```c
#define  RXC    7
#define  TXC    6
#define  UDRE   5
#define  FE     4
#define  DOR    3
#define  OVR    3     /*This definition differs from the ""databook*/
#define  PE     2
#define  U2X    1
#define  MPCM   0
#define UDR (*(volatile unsigned char *)0x2C)

/* SPI */
#define SPCR (*(volatile unsigned char *)0x2D)
#define  SPIE   7
#define  SPE    6
#define  DORD   5
#define  MSTR   4
#define  CPOL   3
#define  CPHA   2
#define  SPR1   1
#define  SPR0   0
#define SPSR (*(volatile unsigned char *)0x2E)
#define  SPIF   7
#define  WCOL   6
#define  SPI2X  0
#define SPDR (*(volatile unsigned char *)0x2F)

/* Port D */
#define PIND  (*(volatile unsigned char *)0x30)
#define DDRD  (*(volatile unsigned char *)0x31)
#define PORTD (*(volatile unsigned char *)0x32)

/* Port C */
#define PINC  (*(volatile unsigned char *)0x33)
#define DDRC  (*(volatile unsigned char *)0x34)
#define PORTC (*(volatile unsigned char *)0x35)
```

```c
/* Port B */
#define PINB (*(volatile unsigned char *)0x36)
#define DDRB (*(volatile unsigned char *)0x37)
#define PORTB (*(volatile unsigned char *)0x38)

/* Port A */
#define PINA (*(volatile unsigned char *)0x39)
#define DDRA (*(volatile unsigned char *)0x3A)
#define PORTA (*(volatile unsigned char *)0x3B)

/* EEPROM */
#define EECR (*(volatile unsigned char *)0x3C)
#define    EERIE    3
#define    EEMWE    2
#define    EEWE     1
#define    EERE     0
#define EEDR (*(volatile unsigned char *)0x3D)
#define EEAR (*(volatile unsigned int *)0x3E)
#define EEARL (*(volatile unsigned char *)0x3E)
#define EEARH (*(volatile unsigned char *)0x3F)

/* Watchdog Timer Control Register */
#define WDTCR (*(volatile unsigned char *)0x41)
#define    WDTOE    4
#define    WDE      3
#define    WDP2     2
#define    WDP1     1
#define    WDP0     0

/* Timer/Counter 2 */
#define ASSR (*(volatile unsigned char *)0x42)
#define    AS2      3
#define    TCN2UB   2
#define    OCR2UB   1
#define    TCR2UB   0
#define OCR2 (*(volatile unsigned char *)0x43)
```

```c
#define TCNT2  (*(volatile unsigned char *)0x44)
#define TCCR2  (*(volatile unsigned char *)0x45)
#define   FOC2    7
#define   WGM20   6
#define   COM21   5
#define   COM20   4
#define   WGM21   3
#define   CS22    2
#define   CS21    1
#define   CS20    0

/* Timer/Counter 1 */
#define ICR1   (*(volatile unsigned int  *)0x46)
#define ICR1L  (*(volatile unsigned char *)0x46)
#define ICR1H  (*(volatile unsigned char *)0x47)
#define OCR1B  (*(volatile unsigned int  *)0x48)
#define OCR1BL (*(volatile unsigned char *)0x48)
#define OCR1BH (*(volatile unsigned char *)0x49)
#define OCR1A  (*(volatile unsigned int  *)0x4A)
#define OCR1AL (*(volatile unsigned char *)0x4A)
#define OCR1AH (*(volatile unsigned char *)0x4B)
#define TCNT1  (*(volatile unsigned int  *)0x4C)
#define TCNT1L (*(volatile unsigned char *)0x4C)
#define TCNT1H (*(volatile unsigned char *)0x4D)
#define TCCR1B (*(volatile unsigned char *)0x4E)
#define   ICNC1   7
#define   ICES1   6
#define   WGM13   4
#define   WGM12   3
#define   CS12    2
#define   CS11    1
#define   CS10    0
#define TCCR1A (*(volatile unsigned char *)0x4F)
#define   COM1A1  7
#define   COM1A0  6
#define   COM1B1  5
```

```
#define   COM1B0    4
#define   FOC1A     3
#define   FOC1B     2
#define   WGM11     1
#define   WGM10     0

/* Special Function I/O register */
#define   SFIOR (*(volatile unsigned char *)0x50)
#define   ADTS2     7
#define   ADTS1     6
#define   ADTS0     5
#define   ADHSM     4
#define   ACME      3
#define   PUD       2
#define   PSR2      1
#define   PSR10     0

/* Oscillator Calibration */
#define   OSCCAL (*(volatile unsigned char *)0x51)
/* On-Chip Debug register */
#define   OCDR (*(volatile unsigned char *)0x51)

/* Timer/Counter 0 */
#define   OCR0 (*(volatile unsigned char *)0x5C)
#define   TCNT0 (*(volatile unsigned char *)0x52)
#define   TCCR0 (*(volatile unsigned char *)0x53)
#define   FOC0      7
#define   WGM00     6
#define   COM01     5
#define   COM00     4
#define   WGM01     3
#define   CS02      2
#define   CS01      1
#define   CS00      0

/* MCU general */
```

```c
#define MCUCSR (*(volatile unsigned char *)0x54)
#define MCUSR  (*(volatile unsigned char *)0x54)
#define     JTD     7
#define     ISC2    6
#define     JTRF    4
#define     WDRF    3
#define     BORF    2
#define     EXTRF   1
#define     PORF    0
#define MCUCR (*(volatile unsigned char *)0x55)
#define     SM2     7
#define     SE      6
#define     SM1     5
#define     SM0     4
#define     ISC11   3
#define     ISC10   2
#define     ISC01   1
#define     ISC00   0

/* SPM Conrol Register */
#define SPMCR (*(volatile unsigned char *)0x57)
#define     SPMIE   7
#define     RWWSB   6
#define     RWWSRE  4
#define     BLBSET  3
#define     PGWRT   2
#define     PGERS   1
#define     SPMEN   0

/* Timer/Counter Interrupts */
#define TIFR (*(volatile unsigned char *)0x58)
#define     OCF2    7
#define     TOV2    6
#define     ICF1    5
#define     OCF1A   4
#define     OCF1B   3
```

```
#define    TOV1      2
#define    OCF0      1
#define    TOV0      0
#define TIMSK (*(volatile unsigned char *)0x59)
#define    OCIE2     7
#define    TOIE2     6
#define    TICIE1    5
#define    OCIE1A    4
#define    OCIE1B    3
#define    TOIE1     2
#define    OCIE0     1
#define    TOIE0     0

/* General Interrupts */
#define GIFR (*(volatile unsigned char *)0x5A)
#define    INTF1     7
#define    INTF0     6
#define    INTF2     5
#define GICR (*(volatile unsigned char *)0x5B)
#define    INT1      7
#define    INT0      6
#define    INT2      5
#define    IVSEL     1
#define    IVCE      0

/* Stack Pointer */
#define SP (*(volatile unsigned int *)0x5D)
#define SPL (*(volatile unsigned char *)0x5D)
#define SPH (*(volatile unsigned char *)0x5E)

/* Status REGister */
#define SREG (*(volatile unsigned char *)0x5F)


/* Port A bits */
#define    PORTA7    7
```

```
#define    PORTA6    6
#define    PORTA5    5
#define    PORTA4    4
#define    PORTA3    3
#define    PORTA2    2
#define    PORTA1    1
#define    PORTA0    0
#define    PA7       7
#define    PA6       6
#define    PA5       5
#define    PA4       4
#define    PA3       3
#define    PA2       2
#define    PA1       1
#define    PA0       0

#define    DDA7      7
#define    DDA6      6
#define    DDA5      5
#define    DDA4      4
#define    DDA3      3
#define    DDA2      2
#define    DDA1      1
#define    DDA0      0

#define    PINA7     7
#define    PINA6     6
#define    PINA5     5
#define    PINA4     4
#define    PINA3     3
#define    PINA2     2
#define    PINA1     1
#define    PINA0     0

/* Port B bits */
#define    PORTB7    7
```

```
#define    PORTB6    6
#define    PORTB5    5
#define    PORTB4    4
#define    PORTB3    3
#define    PORTB2    2
#define    PORTB1    1
#define    PORTB0    0
#define    PB7       7
#define    PB6       6
#define    PB5       5
#define    PB4       4
#define    PB3       3
#define    PB2       2
#define    PB1       1
#define    PB0       0

#define    DDB7      7
#define    DDB6      6
#define    DDB5      5
#define    DDB4      4
#define    DDB3      3
#define    DDB2      2
#define    DDB1      1
#define    DDB0      0

#define    PINB7     7
#define    PINB6     6
#define    PINB5     5
#define    PINB4     4
#define    PINB3     3
#define    PINB2     2
#define    PINB1     1
#define    PINB0     0

/* Port C bits */
#define    PORTC7    7
```

```
#define    PORTC6    6
#define    PORTC5    5
#define    PORTC4    4
#define    PORTC3    3
#define    PORTC2    2
#define    PORTC1    1
#define    PORTC0    0
#define    PC7       7
#define    PC6       6
#define    PC5       5
#define    PC4       4
#define    PC3       3
#define    PC2       2
#define    PC1       1
#define    PC0       0

#define    DDC7      7
#define    DDC6      6
#define    DDC5      5
#define    DDC4      4
#define    DDC3      3
#define    DDC2      2
#define    DDC1      1
#define    DDC0      0

#define    PINC7     7
#define    PINC6     6
#define    PINC5     5
#define    PINC4     4
#define    PINC3     3
#define    PINC2     2
#define    PINC1     1
#define    PINC0     0

/* Port D bits */
#define    PORTD7    7
```

```c
#define   PORTD6    6
#define   PORTD5    5
#define   PORTD4    4
#define   PORTD3    3
#define   PORTD2    2
#define   PORTD1    1
#define   PORTD0    0
#define   PD7       7
#define   PD6       6
#define   PD5       5
#define   PD4       4
#define   PD3       3
#define   PD2       2
#define   PD1       1
#define   PD0       0

#define   DDD7      7
#define   DDD6      6
#define   DDD5      5
#define   DDD4      4
#define   DDD3      3
#define   DDD2      2
#define   DDD1      1
#define   DDD0      0

#define   PIND7     7
#define   PIND6     6
#define   PIND5     5
#define   PIND4     4
#define   PIND3     3
#define   PIND2     2
#define   PIND1     1
#define   PIND0     0

/* Lock,"" and Fuse Bits with LPM/SPM instructions */
```

```
/* lock bits */
#define   BLB12      5
#define   BLB11      4
#define   BLB02      3
#define   BLB01      2
#define   LB2        1
#define   LB1        0

/* fuses low bits */
#define   BODLEVEL   7
#define   BODEN      6
#define   SUT1       5
#define   SUT0       4
#define   CKSEL3     3
#define   CKSEL2     2
#define   CKSEL1     1
#define   CKSEL0     0

/* fuses high bits */
#define   OCDEN      7
#define   JTAGEN     6
#define   SPIEN      5
#define   CKOPT      4
#define   EESAVE     3
#define   BOOTSZ1    2
#define   BOOTSZ0    1
#define   BOOTRST    0


/* Interrupt Vector Numbers */

#define iv_RESET           1
#define iv_INT0            2
#define iv_INT1            3
#define iv_TIMER2_COMP     4
```

```
#define iv_TIMER2_OVF     5
#define iv_TIMER1_CAPT    6
#define iv_TIMER1_COMPA   7
#define iv_TIMER1_COMPB   8
#define iv_TIMER1_OVF     9
#define iv_TIMER0_OVF     10
#define iv_SPI_STC        11
#define iv_USART_RX       12
#define iv_USART_RXC      12
#define iv_USART_DRE      13
#define iv_USART_UDRE     13
#define iv_USART_TX       14
#define iv_USART_TXC      14
#define iv_ADC            15
#define iv_EE_RDY         16
#define iv_EE_READY       16
#define iv_ANA_COMP       17
#define iv_ANALOG_COMP    17
#define iv_TWI            18
#define iv_TWSI           18
#define iv_INT2           19
#define iv_TIMER0_COMP    20
#define iv_SPM_RDY        21
#define iv_SPM_READY      21

/* */

#endif
```

Author Biography

Steven F. Barrett received his bachelor of science degree in electronic enginnering technology from the University of Nebraska at Omaha in 1979, a Master of engineering in electrical engineering from the University of Idaho at Moscow in 1986, and a doctor of philosophy degree from the University of Texas at Autin in 1993. He was formally an active-duty faculty member with the U.S. Air Force Academy, Colorado, and is now an associate professor of electrical and computer engineering in the University of Wyoming. He is a member of IEEE (senior), Tau Beta Pi (chief faculty advisor), and the American Society for Engineering Education. His research interests include digital and analog image processing, computer-assisted laser surgery, and embedded controller systems. He is a registered professional engineer in Wyoming and Colorado. He cowrote with Dr. Pack *Microcontroller Theory and Application: HC12 and S12*, second edition, Pearson Prentice-Hall, 2008; *Embedded Systems Design and Applications with the 68HC12 and HS12*, Pearson Prentice-Hall, 2005; and *Microcontroller Fundamentals for Engineers and Scientists*, Morgan and Claypool, 2006. In 2004, Barrett was named Wyoming Professor of the Year by the Carnegie Foundation for the Advancement of Teaching. E-mail: steveb@uwyo.edu.

Daniel J. Pack is a professor in the Department of Electrical Engineering at the U.S. Air Force Academy, Colorado. He received his bachelor of science degree in electrical engineering in 1988 from Arizona State University, a master of science degree in engineering sciences in 1990 from Harvard University, and a doctor of philosophy degree in electrical engineering in 1995 from Purdue University. He was a visiting scholar at Massachusetts Institute of Technology–Lincoln Laboratory. He coauthored five textbooks on microcontrollers and embedded systems and authored more than 70 journal and conference papers. He is a member of Eta Kappa Nu, Tau Beta Pi (faculty advisor), IEEE (senior), and American Society for Engineering Education. He is a registered professional engineer in Colorado. In 2005, Dr. Pack was named Colorado Professor of the Year by the Carnegie Foundation for the Advancement of Teaching. His research interests include cooperative UAVs, intelligent control, automatic target recognition, and robotics. E-mail: daniel.pack@usafa.edu.

Index

AC device control, 114
AC interfacing, 114
ADC, 35
ADC conversion, 35
ADC process, 43
analog sensor, 104
analog vs digital signal, 36
ASCII, 23
ATmega16 ADC, 46
ATmega16 interrupt system, 57
ATmega16 timers, 74
ATmega16 timing system, 67

Baud rate, 22
Bell Laboratory, 38

counter-based ADC, 46
counting events, 72
crystal time base, 68
CTC timer mode, 78
current sink, 97
current source, 97

DAC converter, 53
data rate, 42
DC motor, 114
DC motor control, 112
DC motor speed control, 73
decibel (dB), 42

digital sensor, 101
dot matrix display, 107
duty cycle, 68
dynamic range, 42

electrical specifications, 96
encoding, 38, 40

fast PWM timer mode, 78
Flash EEPROM, 4
flight simulator panel, 115
frequency measurement, 72
full duplex, 22

Harry Nyquist, 38
HC CMOS, 96

ImageCraft ICC AVR compiler, 60
input capture, 71
input capture programming, 91
input devices, 97
integration ADC, 46
internal interrupt, 62
interrupt theory, 57

keypad, 101

LED biasing, 105
light emitting diode (LED), 105

liquid crystal display (LCD), 107

MAX232, 23
MOSFET, 112
motor operating parameters, 114

normal timer mode, 77
NRZ format, 22
Nyquist sampling rate, 38

operating parameters, 95
output compare, 73
output device, 105
output timer, 70

parallel ADC, 46
parity, 23
period, 68
phase correct timermode, 78
photodiode, 44
PWM, 73
PWM programming, 90

quantization, 38

RAM, 4
resolution, 41
RS-232, 23

sampling, 38
sensors, 101
serial communications, 21
servo motor, 114
signal conditioning, 43
signal generation, 73
solid state relay (SSR), 112

SPI, 29
stepper motor, 114
successive-approximation ADC, 45
switch debouncing, 99
switch interface, 99
switches, 97

time base, 7
Timer 0, 75
Timer 1, 81
Timer 2, 85
timer applications, 71
timer modes, 77
timer system programming, 87
timing system, 67
transducer interface, 43
tri-state LED indicator, 107
TWI, 33

USART, 23

volatile, 4

Made in the USA
Lexington, KY
29 October 2011